Pesticides: Boon or Bane?

M. B. GREEN

In developing countries, the use of pesticides makes possible the production of adequate amounts of foods with satisfactory nutritional value and at reasonable prices. The pesticides also alleviate the problems of scarce agricultural land and produce certain savings in labor and energy. In addition, they are of great help in protecting public health and, specifically, in the control of noxious insects. Unfortunately, this is not the whole story. Benefits are apparently not realized without the very serious possibility of hazards and dangers. The almost unbounded enthusiasm that accompanied the initial massive world-wide introduction of pesticides has, in recent years, changed to skepticism, criticism, and fairly deep distrust. This attitude is not only a public opinion phenomenon but is also shared by considerable segments of the scientific community. How much of this is justified?

Clearly the paradise originally promised is not realistic. Some pesticides can attack non-targets as well as target species. Some might enter into food chains and harm birds and small mammals. Some might damage fish and aquatic life by runoff from agricultural land into rivers and streams. Some might leave unacceptably large residues in food crops which could present a toxic hazard to the consumer. How real and how great are these dangers? Has there really been enough research and is there constructive evidence clearly defining the kind and degree of harm?

This book examines both the benefits and the risks and weighs them against each other. The value of pesticides to the national economy and well-being is measured against the risks of possible harm to the environment. The author advocates the use of cost benefit analysis to determine what restrictions or prohibitions should be placed on the use of a particular pesticide. But he also points out that the result of a cost benefit analysis for a particular pesticide in one country may not be the same as that in another country and that the factors of the equation might be totally different if the food and health needs of the developing countries are considered.

Dr Maurice Berkeley Green directs a large unit in the Research Department of Mond Division, Imperial Chemical Industries Ltd, with special responsibilities for the development of new crop protection products. He also teaches technological economics of crop protection at the University of Stirling, Scotland.

WESTVIEW ENVIRONMENTAL STUDIES
Editors: J. Rose (UK) and E. W. Weidner (USA)

WESTVIEW ENVIRONMENTAL STUDIES

Editors: J. Rose (UK) and E. W. Weidner (US)

PESTICIDES— BOON OR BANE?

M.B. Green

WESTVIEW PRESS • BOULDER • COLORADO

Westview Environmental Studies: Volume 1

*Copyright © 1976 in London, England
by Elek Books Ltd.*

*Published 1976 in London, England
by Paul Elek Ltd.*

*Published 1976 in the United States of America by
    Westview Press, Inc.
    1898 Flatiron Court
    Boulder, Colorado 80301
    Frederick A. Praeger, Publisher and Editorial Director*

Printed and bound in Great Britain

Library of Congress Cataloging in Publication Data

Green, Maurice Berkeley.
  Pesticides.

  (Westview environmental studies ; v. 1)
  Bibliography: p.
  Includes index.
  1. Pesticides. 2. Pesticides—Environmental
aspects. I. Title.
SB951.G76   632'.95   76-5881
ISBN 0-89158-610-5

TO LUCY

And he gave it for his opinion, that whoever could make two ears of corn or two blades of grass to grow upon a spot of ground where only one grew before, would deserve better of mankind, and do more essential service to his country, than the whole race of politicians put together.

JONATHAN SWIFT, 1667–1745

Concern for man himself and his fate must always form the chief interest of all technical endeavour.

ALBERT EINSTEIN, 1879–1955

Contents

Preface

In recent years, a great deal has been written and spoken about the alleged hazards of pesticides, but the benefits of pesticides have received little publicity. The purpose of this book is to examine both the benefits and the risks objectively, and to weigh them against each other. It is hoped that it will assist members of the public to form balanced opinions on the subject and enable them to influence legislators in Governments to take decisions which will result in pesticides being used in the most effective and least harmful way for the well-being of the whole community, so that both the nation's food supplies and the nation's environment are safeguarded.

It has proved very difficult to discuss all the many complexities of crop protection and pest control within the constraints of space of this short book. The author, therefore, readily pleads guilty in advance to the criticism that he has dealt with some topics very superficially. There is, however, a list at the end, of selected books for further reading to help those who are interested to gain more knowledge and understanding.

The author acknowledges with gratitude the many books and papers on which he has drawn for source material. He wishes particularly to thank Dr J. M. Barnes, Director of the Toxicology Research Unit of the UK Medical Research Council; Dr R. D. Bowden, Mond Division, Imperial Chemical Industries Limited; Professor F. R. Bradbury, Department of Industrial Science, University of Stirling; Dr W. B. Ennis and his colleagues, United States Department of Agriculture, Beltsville; Professor F. G. W. Jones, Deputy Director of Rothamsted Experimental Station; Dr K. Mellanby, Director of Monks Wood Experimental Station of the UK Nature Conservancy; Mr A. H. Strickland, Plant Pathology Laboratory, UK Ministry of Agriculture, Fisheries and Food; Mr G. A. Wheatley, UK National Vegetable Research Station and Mr C. J. Lewis and Mr J. M. Winchester, Plant Protection Limited, for helpful advice and criticism.

The author is especially grateful to Miss V. Weston who cheerfully and competently typed the manuscript from almost illegible handwriting.

The opinions expressed in this book are entirely those of the author and do not necessarily reflect the views of Imperial Chemical Industries Limited.

Glossary

The word 'pesticide' is an ugly word but it has now become established in general use and is not likely to be replaced. There is a need for an omnibus term to describe chemicals which are used to control any pests or diseases which attack crops, animals, men or materials, and nobody has suggested an acceptable alternative. It is important to realise that the word 'pesticide' does not, as is often thought, apply only to chemicals used to control noxious insects and fungi, but also that it includes weedkillers, since weeds are pests.

The term 'crop protection chemical' applies only to substances used to protect growing crops. The term 'agricultural chemical' or 'agrochemical' includes not only crop protection chemicals but also fertilisers and other plant nutrients. The term 'economic poison' applies only to insecticides. The term 'pest control chemical' is usually restricted to substances used to control pests of animals (such as ticks and blowflies), pests of stored products (such as rats and mice) and pests which affect public health (such as cockroaches, bedbugs and mosquitoes). The term 'industrial biocide' is applied to substances used to control living organisms which attack materials (such as fungi on paint or termites in wood). The term 'phytopharmaceutical' applies to chemicals used to control the diseases rather than the pests of plants.

The following are the chief types of pesticides:

Herbicides	Commonly known as weedkillers and used to control unwanted plants either in crop or in industrial and amenity situations. They can be selective (i.e. kill the weeds but not the crop) or general (i.e. kill all vegetation). A substance which kills a plant is said to be phytotoxic.
Sylvanicides	Herbicides used for killing trees and bushy plants.
Rodenticides	Used to control rats and mice.
Avicides	Used to control bird pests, especially in tropical areas.
Insecticides	Used to control insects in crops, on animals and in public health. They can be contact (i.e. kill the insect directly) or systemic (i.e. make the plant poisonous to the insect).
Larvicides	Used to kill the larval stages of an insect.

Ovicides	Used to kill insect eggs.
Ixodicides or *Tickicides*	Used to control ticks, which mainly infest animals.
Miticides	Used to control mites, which differ from insects in many ways, e.g. they have eight legs, and unsegmented bodies.
Acaricides	Used to control mites which belong to the family of small spiders and mainly infest fruit trees.
Nematicides	Used to control the nematodes, which are microscopic worms which inhabit the soil and mostly attack plants via the roots.
Fungicides or *Antifungals*	Used to control the fungi which attack plants and include the mildews, blights and wilts. Can be protective (i.e. kill the fungal spores alighting on the plant) or systemic (i.e. make the whole plant lethal to the fungi).
Bactericides or *Antibacterials*	Used to control bacteria, commonly known as germs, which affect plants, animals and humans.
Antivirals	Used to control the viruses, which are the smallest micro-organisms, responsible for a wide range of lethal and sub-lethal diseases in plants, animals and humans.

Throughout this book SI metric units are used. For those not familiar with these one hectare (ha) = about 2·5 acres, one kilogram (kg) = about 2·25 pounds, one tonne (t) = about 1 ton, one litre (l) = about 1·75 pints, one metre (m) = about one yard. The currency used is the USA dollar equal to about 0·45 pounds Sterling. The USA billion equal to one thousand million (10^9) is used.

1

Historical Perspective

A contrast is often drawn between the artificial, hurried, neurotic life of the industrial city-dweller and the natural, calm well-balanced life of the countryman living in age-old harmony with the soil. Yet man is, by origin, a nomadic, predatory hunter, possibly better equipped genetically for the cut and thrust of life in the urban jungle than for a static rustic existence. Cowper said, 'God made the country and man made the town', but the fact is that there is nothing 'natural' about the practices of horticulture and agriculture, and the cultivated countryside as we see it today bears no resemblance to the appearance and character it would have had if man had not deliberately modified his environment and adapted nature to his own purposes. *Homo sapiens* is a physically weak species which has survived and prospered by virtue of an intelligence which enabled him to manipulate and make advantageous changes in his environment, and the change which was most far-reaching and which had eventually the greatest environmental impact was when he decided to settle in communities in one place, since he discovered in these how to cultivate plants to feed himself and his animals, rather than to move about constantly seeking new food supplies for himself and new pastures for his flocks. In this way he released himself from the necessity of a continual wandering search for sustenance and shelter and gave himself a reasonable assurance of sufficient to eat and enough leisure to develop his culture. Only then could human progress become possible, and the complex of ideas and institutions which we call civilisation evolve.

Development of agriculture and horticulture could not take place as long as all men were nomadic, because wild animals do not lend themselves easily to domestication and wild plants give poor yields—for example, compare the wild oat with the cultivated varieties. It started in those primitive townships of some 10 000 years ago where communities had already established themselves permanently in one place by trading materials, such as metals for weapons or pigments for self-decoration, with the surrounding hunting tribes in exchange for supplies of food. At first, such supplies comprised wild animals and edible seeds, but the need to maintain supplies of fresh meat in the townships led to development of stock breeding and raising. Planting of mixed seeds brought plants from widely different localities into close contact for the first time and led, by chance, to higher yielding varieties which man, with his intelligence, later systematically exploited, thus developing horticulture. Wheat and barley emerged in the Middle East, rice in eastern Asia and maize in Central America, all areas shown archaeologically to be amongst the earliest sites of established communities.

Agriculture and horticulture—the technologies of interfering with nature—were, therefore, inventions made by man in the towns and practised there long before they spread to the countryside. This happened when the townships grew larger and more space was needed for animals and plants. By 6000 BC some of the townships were quite sizeable; for instance, the settlement excavated by Mellaart at Catal Hüyük in Anatolia had houses covering fourteen hectares and a population of several thousands. The practices and people of the town were transplanted to the country and the nomadic peoples either settled into agricultural villages set up by townsfolk to serve the towns or were driven away. This process is discussed in detail in Jane Jacob's book *The Economy of Cities*. Since that time, agriculture and horticulture have continually advanced by means of the technology which man could inject into the countryside from the towns and cities, ranging through the three-field system of the twelfth century and the crop rotation of the eighteenth century to the modern techniques of mechanised farming equipment. This is why the most agriculturally advanced and productive nations in the world today are those that have developed the most highly industrialised cities. The great increase in productivity of Japanese agriculture since World War II is an example from recent history of the impact of urban technology, but development of agriculture there lagged behind that of industry just as Adam Smith in *The Wealth of Nations* tells us it did in England in the eighteenth century.

The displacement of nomadic tribes by agriculture and horticulture introduced from the towns round about 6000 BC is probably the underlying meaning of the great contemporary allegory of Cain and Abel. Cain was a tiller of the fields and Abel was a nomadic keeper of sheep. The murder of Abel symbolises the replacement of one way of life by the other, but, as we know from Genesis, there is a price to be paid for this.

Nature aims constantly at a varied flora, as anyone who has left his garden untended for several weeks during a vacation realises only too clearly. Monoculture, which is the cultivation of serried rows of the same plant over a wide area, can be maintained only by a never-ending struggle with the environment to remove competing plants, which spring from seeds which are already present in the soil or which are introduced to it either through the air or as impurities in the crop seed. The extent of Nature's opposition to monoculture should not be underestimated: each square metre (about 1 square yard) of average soil in the UK contains about 40 000 weed seeds, and one plant of, for example, hedge mustard disperses into the air about 750 000 seeds during one season. If all these seeds germinated and produced plants which each seeded to the same extent, the third generation of plants would more than cover the total area of the earth's surface. There are about 30 000 identified weed species in the world and, of these, about 1800 can cause serious economic losses.

Weeds generally do not damage crop plants directly but, by competing with them for available water and nutrients in the soil and by obscuring light they restrict crop growth and reduce crop yields. For the peasant farmer in developing countries such losses may mean starvation for his family, while, for the highly organised farmer in developed countries, they may result in severe financial loss. Some weeds are truly parasitic and some cause delayed damage by acting as alternative or overwintering hosts for pests and diseases. Some also produce exudates toxic to other plants.

Weeds are, however, by no means the only problem with which man has had to grapple in developing and sustaining horticulture and agriculture. A host of other living organisms attack and devour his crops, both before and after harvest, and carry disease and death to his plants and animals and to man himself. The human race has maintained a precarious foothold on this planet by a stark and unrelenting war against voracious and ruthless natural enemies which, over the centuries, have taken a far greater total of human life than all the internecine wars of man put together.

Amongst these enemies are the birds and small mammals which take their share of seed, growing crops and stored produce. The tradition in sowing used to be expressed as 'One for the rook, one for the crow, one for to rot and one for to grow'. All animals, including man, are dependent, directly or indirectly, on the organic food which plants synthesise from carbon dioxide and water by the action of light on the green leaf, so, naturally, they compete for edible food. The effects of a species of animal introduced into an area where there is ample food and few natural enemies can be seen in the ravages of the rabbit in Australia. In tropical Africa, flocks of weaver birds can approach locust swarms in size and do similar damage, and enormous flocks of starlings have, at times, ravaged the Middle West of the USA. The world-wide depredations of rats and mice are well known, and it is the rat flea which carries bubonic plague, the 'Black Death'.

Insects, mites and nematodes are major pests which have a 200 million year evolutionary edge over mankind and which attack both animals and growing and stored crops. About 10 000 of the 700 000 known species of insects are recognised as pests and about 30% of the food grown in the world still goes to feed them rather than to feed people. There are about 15 000 species of nematodes of which about 1500 can cause serious damage to plants. In addition, the insects and mites which infest animals

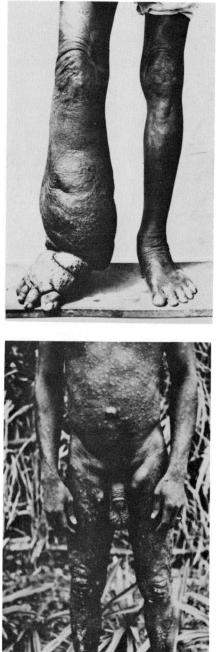

PLATE 1 Infection with filariasis —an insect-borne disease. (By courtesy of Wellcome Museum of Medical Science, England.)

PLATE 2 Infection with onchocerciasis—an insect-borne disease. (By courtesy of Wellcome Museum of Medical Science, England.)

PLATE 3   Infection with trypanosomiasis—an insect-borne disease. (By courtesy of Wellcome Museum of Medical Science, England.)

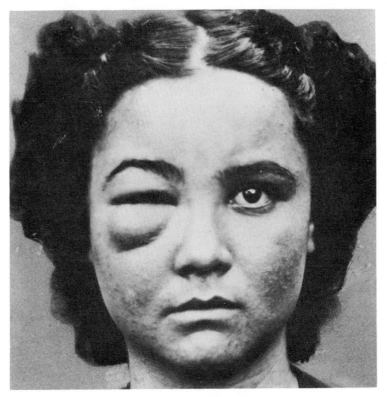

PLATE 4   Infection with Chaga's disease—an insect-borne disease. (By courtesy of Wellcome Museum of Medical Science, England.)

PLATE 5 Sheep infested with scab. (By courtesy of Wellcome Veterinary Research Laboratory, England.)

PLATE 6 Jersey cow infested with cattle tick. (By courtesy of Wellcome Veterinary Research Laboratory, England.)

and humans carry and transmit some of the most debilitating diseases which mankind has to endure, such as malaria, sleeping sickness, yellow fever, plague, bilharzia and many others (see Plates 1, 2, 3, 4). The magnitude of the suffering caused is immense: prior to the WHO malaria eradication programme, it was estimated that there were about 100 million people in the world suffering from this disease. The effects of infestations on animals, for example, ticks on cattle or blowfly on sheep, are hideous (see Plates 5, 6).

Amongst the lower organisms which are enemies of man are the fungi—the mildews, blights and wilts—and the bacteria and viruses. There are about 50 000 species of fungus now recognised in the world of which about 1500 can cause severe disease to plants. The fungi, by and large, attack mainly growing plants and stored products and some of their names—such as take-all of wheat—are very descriptive of their effects. Even in recent history, potato blight has caused two major famines in Ireland. Moses specified attacks of mildew as one of the punishments for failure to obey the Lord's commandments. Even when fungi do not kill plants they can produce serious reductions in crop yields, for example, it is estimated that the incidence of mildew on barley causes an average reduction of about 14% in yields of this cereal in the UK. Bacteria more commonly affect warm-blooded creatures, such as men and animals. Their effects are so well known that there is little need to give illustrative examples. The viruses are ubiquitous and invade men, animals and plants and commonly, in sub-lethal infections, retard growth and development.

It is not possible to estimate precisely the total losses of food and fibre which are caused all over the world by these enemies of crops, but they are certainly prodigious. It was estimated that, in the USA, between 1951 and 1960 annual losses of potential agricultural production due to various causes were the following percentages of total production by weight: by weeds, 9%; by birds and small mammals, 2%; by insects and mites, 13%; by nematodes, 1%; by disease, 10%; overall total, 35%. A more recent estimate gave the crop losses in the USA in 1967 in cash terms as $5.36 billion by weeds, $5·07 billion by insects and mites, $0·4 billion by nematodes and $3·66 billion by diseases, totalling $14·50 billion compared with a total value of actual crop production in the USA of $21·4 billion. Attacks by these various pests and diseases may occur successively so that their effects are multiplied rather than added and so may result in total loss of a crop.

The Food and Agricultural Organisation of the United Nations has estimated that the following percentages of total potential world production are lost to pests and diseases: wheat, 35%; potatoes, 40%; sugar beet, 24%; apples, 30%; cotton, 60%; tobacco, 62%. If 35% of total potential crop production can be lost in a highly advanced country such as the USA which has the most modern technology at its command, it can be imagined what the losses must be in developing countries. The World Health Organisation has hazarded an informed guess that, in many of these countries twice as much is lost as is finally harvested. These figures have to be viewed against a background of the problems of feeding a world population which, at the present rate of increase, is doubling itself every thirty-seven years. In India, for example, during the past ten years, the population has increased by 22%, but total food production in that country has, during that period, increased by only 0·4%.

The question may be asked why the practices of horticulture and agriculture so greatly exacerbate the effects of pests and diseases. The answer lies in the fact that most insects, fungi and bacteria have evolved specifically to live on one source of food and to infest one particular type of plant or animal. In a natural environment, any plant which can be a host for one organism will be surrounded by other types of plants which cannot and so, for any particular organism, the supply of food will be limited and this, together with the presence of predators—natural enemies of the organisms—limits reproduction and population growth. This is, essentially, what is meant by the balance of nature. A large area covered by plants of the same kind standing close together offers abundant food and no barriers to inhibit movement of any pest or disease for which that type of plant is a host. The attacks can therefore build up exponentially throughout the whole crop. The same is essentially true for large stores of a commodity such as grain, the existence of which can enormously increase rodent populations, or for large herds of animals or massive populations of men jammed together in cities—conditions under which infestations and epidemics can run riot.

The foregoing is an oversimplification, as population dynamics are a very complex problem which cannot be discussed in detail in this book. Nevertheless, it is clear that natural communities of animals, plants, insects and fungi have a much greater stability than the artificial communities produced by agriculture and horticulture. In them, there are rarely enough of any particular pest present to destroy the host plants and, in general, population numbers are considerably below those that the host plants could carry. The artificial communities are inherently unstable and are prone to epidemics. The reason appears to be that when man destroyed the natural vegetation as a prerequisite to agriculture and horticulture, he disrupted the complex interrelationships which had grown up over long periods between plants and living creatures, and few of the latter survived this change. The surviving species generally failed to fill all the niches provided by the new crops and, as Nature abhors a vacuum, other species moved in. So, it was only a matter of time before a potentially harmful species from one place found that some conditions ideally suited to its way of life had been created somewhere else by the new practices of agriculture and horticulture. This process has been speeded up by modern transport systems which have distributed harmful, or potentially harmful, species all over the world.

During his history, man has used his developing technological skills to assist him in his fight against pests and diseases. Against weeds, the two main methods of defence have been, firstly, to plough and till the soil so that unwanted seeds and plants are disturbed and a clean surface presented for sowing, which allows the crops and weeds to at least start on equal terms, and, secondly, to remove weeds as they grow by hand-picking and hoeing. As technology has developed, wooden ploughs and hoes drawn by horses or other animals have been replaced by metal implements and then by all the array of mechanical equipment of the modern farm. Against birds and small mammals man has relied for protection on traps and scaring devices and, later, guns, wire fences and poison baits. A novel method of rodent control was tried in Hamelin but brought disastrous consequences! Against insects and fungi man, throughout his history, has had few defences and the stories of the Old Testament bring home to us how ever present the threat of plagues of insects or diseases and consequent famine have been. Even the Romans, who laid the foundations of so much other technology, could do no better than hold an annual rite—the Robigalia—to try to protect their cereals from mildew and rust, but the efficiency of this method of crop protection is doubtful. The main defences have been destruction of any waste which could be a repository for pests or diseases and rotation of crops to prevent population build-ups.

The possibility of using chemical substances in the fight against pests and diseases occurred quite early to man. A papyrus of 1500 BC records formulae for preparation of insecticides against lice, fleas and wasps. One of the main sources of man's interest in chemistry was his search for means to relieve human ailments, and every primitive society had its store of folklore remedies, often jealously guarded by a priestly élite. These remedies were based on naturally occurring organic substances, such as the juice of certain plants, or on inorganic substances, such as powdered minerals. It was natural, therefore, for man also to seek similar remedies for the pests and diseases of his plants and animals but, unfortunately, he was not very successful.

One of the first successful attempts to use a deliberately formulated mixture of mineral substances to control a plant disease came during the industrial and scientific revolution of the nineteenth century when, in 1882, Millardet demonstrated the value of Bordeaux mixture as a fungicide. In 1867 farmers in the United States, faced with an invasion of Colorado beetle into cultivated potato lands, laid the foundations of chemical insecticides by applying an arsenical poison, Paris Green, to their crops. The observation in 1897 by a French vine-grower, Bonnet, that charlock was killed by application of a solution of copper sulphate, whereas oats in the same environment were unaffected, opened up the concept of selective herbicides—that is, chemicals which could kill weeds but leave crop plants undamaged. In 1901, Bolley in North Dakota discovered the similar effect of ferrous sulphate, which was subsequently used for many years in the US for control of broad-leaved weeds in cereal crops. Thus, in the second half of the nineteenth century the foundations of the three main groups of modern pesticides, insecticides, fungicides and selective herbicides, were laid (see Plates 7, 8, 9, 10, 11, 12).

During the early years of the twentieth century other chemical crop protection agents were brought into use. These were mainly inorganic substances, such as polysulphides and other derivatives of sulphur, arsenicals and compounds of metals such as lead, copper and mercury. In addition there were a few substances derived from natural sources such as nicotine, pyrethrum, derris and quassia, together with some crude distillates from coal tar.

From the middle of the nineteenth century, organic chemistry, that is the chemistry of compounds of carbon, hydrogen and oxygen similar to those derived from natural sources, had made great strides. This knowledge was first applied in the late nineteenth century to development and production of dyestuffs and then, from the beginning of the twentieth century to discover and develop new synthetic organic chemicals which would be useful for treatment of human diseases. By World War II, a flourishing pharmaceutical industry had been built up, in many cases as a diversification by companies producing dyestuffs. Attempts were made to apply this type of chemistry to discover synthetic organic compounds to combat pests and diseases of plants and animals, but it was not until World War II that any substantial advances were made. It was then that three major break-throughs came, which started the great developments in pesticides which have taken place during the past thirty years. These were the discovery of the insecticide DDT, the discovery of the organophosphorus insecticides, and the discovery of the selective phenoxyacetic herbicides, such as 2,4-D and MCPA.

Since that time, a whole range of synthetic products have been discovered and developed and about 6000 novel compounds have been patented, of which about 600 have found some commercial use. It is a sad reflection on human nature that the technology of homicides is as old as mankind, stretching from the curare-tipped dart of the primitive Amazonian Indian to the megaton nuclear device of his more civilised counterparts, whereas the technology of pesticides goes back hardly a century. Moreover, although DDT and the organophosphorus compounds were developed originally from a specific search for

PLATE 7   Wild oat in barley—untreated. (By courtesy of Plant Protection Limited, England.)

PLATE 8   Barley treated with a selective herbicide. (By courtesy of Plant Protection Limited, England.)

PLATE 9 Barley affected by mildew—untreated. (By courtesy of Plant Protection Limited, England.)

PLATE 10 Barley treated with a systemic fungicide. (By courtesy of Plant Protection Limited, England.)

PLATE 11 Sugar beet infested with aphis—untreated. (By courtesy of Plant Protection Limited, England.)

PLATE 12 Sugar beet treated with an insecticide. (By courtesy of Plant Protection Limited, England.)

insecticides by Müller in Switzerland and Schrader in Germany, respectively, all research on the latter group of compounds was switched to develop 'nerve gases' for use in war, when it was realised that they had potential for this purpose, although, thankfully, they were never actually used. The selective phenoxyacetic herbicides were developed not as a result of research aimed at assisting horticulture, but from a secret wartime project to develop compounds which could destroy rice fields and other vegetation and might thus be used as a weapon to starve enemy populations into submission.

2

Benefits and Risks

The impact of modern crop protection chemicals, fertilisers and mechanical equipment on agricultural technology and practices in the developed countries during the past thirty years has been little short of revolutionary. Yields per hectare of all major crops have increased steadily year by year as have total outputs of all food crops (Tables 2.1 and 2.2). These increases have been achieved with a steady reduction in

TABLE 2.1

Increase in crop yields.

(a) Great Britain (100 kg/ha). (From *A Century of Agricultural Statistics*, HMSO, London.)

Crop	1915	1925	1935	1945	1950	1955	1960	1965
Wheat	22.0	23.0	23.3	24.0	26.3	33.5	35.8	40.7
Barley	18.0	19.7	21.2	23.9	24.1	32.1	33.0	37.7
Oats	17.7	18.8	20.3	21.7	21.7	26.9	26.6	31.5
Potatoes	158.1	165.7	158.1	180.7	193.3	183.2	218.4	256.0
Sugar beet	—	215.9	228.4	236.0	271.1	308.7	419.2	374.0
Turnips and swedes	358.9	333.8	306.2	379.0	389.1	364.0	484.4	494.5

(b) USA (100 kg/ha). (From *Agricultural Statistics*, USDA, Washington.)

Crop	1938	1948	1958	1968
Wheat	8.4	11.4	17.4	18.0
Corn	16.3	25.0	30.3	46.1
Cotton	2.6	3.4	5.1	5.6

TABLE 2.2

Increase in total output of crops in Great Britain (millions of tonnes). (From *A Century of Agricultural Statistics*, HMSO, London.)

Crop	1915	1925	1935	1945	1950	1955	1960	1965
Wheat	1.9	1.4	1.7	2.1	2.6	2.6	2.9	4.0
Barley	0.9	1.1	0.7	2.0	1.9	2.9	4.1	7.8
Sugar beet	—	1.1	3.4	3.8	5.2	4.5	5.9	6.7

the labour forces required (Table 2.3), and it is just as well that this was possible, because there has been a continuous drift of people away from the countryside into the towns. The practice of crop rotation has diminished and continuous monoculture has become much more common. More and more farmers have tended to become specialists in

TABLE 2.3
Labour in agriculture
(a) Great Britain. (From *A Century of Agricultural Statistics*, HMSO, London.)

Year	Thousands engaged in agriculture	Average weekly wages in agriculture $/week
1925	925	3·3
1935	787	3·6
1945	887	8·0
1950	843	11·1
1955	732	14·7
1960	645	18·6
1965	514	23·1

(b) USA. (From *Agricultural Statistics*, USDA, Washington.)

Year	% of economically active population in agriculture	Average monthly wages in agriculture $/month
1940	18·7	37
1950	12·2	121
1960	7·8	192
1965	5·9	240

one crop and the average size of farms has increased (Table 2.4). Thirty years ago, one US agricultural worker produced, on average, enough food and fibre for his own needs and those of nine other people. Today, one farm worker in the USA produces enough for himself and thirty-one others. One-fifteenth of the total population of the USA working on only half as many farms as thirty years ago feed a population that has grown in that time by 50%. The people of the USA are better fed than they have ever been, yet they spend an average of only 19% of their take-home pay on food. These levels of efficiency of production and reasonable prices could not have been achieved and cannot be maintained without the use of chemical crop protection techniques.

A number of studies have been carried out, particularly by the Department of Agriculture in the USA, to assess what the effect would be if certain of the most commonly used crop protection chemicals were

TABLE 2.4
Change of size in farms over 50 years
(a) England and Wales (hundreds). (From *A Century of Agricultural Statistics*,
HMSO, London.)

Year	< 40 ha	40–120 ha	120–200 ha	200–400 ha	> 400 ha
1915	3493	696·8	—	—	—
1925	3257	672·9	—	—	—
1935	3015	663·2	89·3	26·9	3·1
1945	2825	654·2	89·3	29·1	4·3
1955	2925	640·3	93·5	31·7	5·4
1960	2685	626·2	96·4	35·6	6·5
1965	2380	584·2	103·3	44·6	9·3

(b) USA (thousands). (From *Historical Statistics*, US Dept of Commerce, Washington.)

Year	< 40 ha	40–100 ha	100–200 ha	200–400 ha	> 400 ha
1900	3297	1912	378	103	47
1910	3692	2051	444	125	50
1920	3775	1980	476	150	67
1930	3733	1864	451	160	81
1940	3577	1796	459	164	101
1950	3011	1590	478	182	121
1954	2560	1471	482	192	131

(c) USA. (From *Agricultural Statistics*, USDA, Washington.)

Year	Number of farms millions	Total land in farms millions ha	Average land per farm ha
1960	3·96	476	120
1965	3·36	461	138
1970	2·95	446	151
1973	2·83	441	156

withdrawn. For example, a report in 1970 estimated that prohibition in the USA of the use of the phenoxyacetic herbicides 2,4-D and 2,4,5-T would increase the direct costs of USA farm production by about $290 million per year and would necessitate 20 million more hours' work, without additional income, from farmers and their families to maintain current levels of crop production. It has been estimated by the Food and Agricultural Organisation of the United Nations that cessation of all use of crop protection chemicals in the USA would reduce total output of crops and livestock by 30% and would increase the price of farm products to the consumer by between 50% and 70%.

This has been a quiet revolution which has largely escaped notice by the public, but the results of which they now take for granted. The consumer currently demands a consistent quality and appearance in foodstuffs which was unapproachable fifty years ago. To find an occasional maggot on biting into an apple used to be an accepted hazard, but such an occurrence in a polythene-wrapped pack of evenly-sized, completely unblemished, apples from a supermarket would nowadays provoke an immediate complaint. To come across a pod full of grubs when shelling peas was, at one time, a normal event but now one grub in a pack of frozen peas would precipitate a visit to the local inspector of food, and the discovery of insect contamination in a tin of baby food could, if publicised, seriously damage a manufacturer's reputation and business. Good appearance and substantial freedom from blemish or damage— which is referred to as zero tolerance of pest and disease damage—are a prerequisite nowadays for a grower to obtain a reasonable price for his crops, and they can be assured economically only by intensive use of crop protection chemicals.

From the point of view of the grower, crop protection chemicals, when properly used, have proved to be a very rewarding investment. Many studies of this type have been made. For example, it was estimated that, in 1963, an expenditure of one dollar on pesticides brought an average increase of four dollars in the value of produce sold from the farm. A series of experiments with potatoes over a period of ten years indicated $6.71 average return for each $1 expenditure on pesticides, and a similar study on apples showed $5.17 average for each $1 expended. It has been estimated that the increase in value of crops harvested on ten million hectares in Canada treated with herbicides in 1960 at a cost of $8 million was $58 million—equivalent to a return of 7 to 1. Studies on German farms growing cereals and root crops over a period of four years showed a gain in crop value of $47.2 per hectare for an outlay in crop protection chemicals of $20.3 per hectare. The returns obtained from use of pesticides will vary from crop to crop and from locality to locality and are greatly influenced by the skill and efficiency with which the farmer uses crop protection methods. Nevertheless, there is ample evidence that, for agricultural production as a whole, use of pesticides has proved very rewarding financially both to the farmer and to the consumer.

The need for pesticides extends beyond their role in crop protection. Substantial livestock and poultry losses occur as a result of diseases, insects and internal parasites. Losses of forest resources and products caused by pests and diseases are considerable. Forage production on rangeland can be increased two to four times by controlling brush with herbicides. Insecticides have greatly increased public health by controlling common pests such as fleas, cockroaches, lice and bedbugs and by combating the insects which carry a variety of diseases, particularly in tropical areas. Malaria has been eradicated in the USA and seventeen other countries and is well under control in thirty more. Not only has this antimalarial campaign enormously decreased human suffering, but it has also opened up vast areas of formerly uninhabitable lands. Herbicides have an important part to play in the control of weeds on industrial sites, in towns and cities, on roadside verges and in amenity and recreational areas of land and water. Living organisms cause great losses by attacking materials, as well as crops and animals. For example, termites devour wood, seaweeds foul ships, fungi rot textiles and wood and spoil paintwork and works of art, rodents gnaw electricity cables, birds foul buildings, effects which are known collectively as 'biodeterioration'.

It is reasonable to ask at this stage whether use of crop protection and pest control chemicals has any disadvantages. Is there a price to be paid? These chemicals have been deliberately developed to be toxic to some living organisms and this is the reason for their commercial utility. There is a unity between all forms of life, so accidental ingestion of pesticides by humans or animals might produce adverse effects if they were very poisonous. In this case, there would be a possibility of health risks to the operatives who are actually engaged in handling and spraying them. There is the possibility of hazard to children, when the carelessness of adults permits them access to such chemicals, and the possibility of their being deliberately used for suicide or murder. Since crop protection chemicals are applied to plants which will produce edible crops and, in some cases, to the crops themselves, there is the possibility that small residues of the chemicals might remain in the crop until it is eaten by the consumer and that, if this happens, it might be deleterious to the consumer's health either acutely or in the long term. Furthermore, since crop protection chemicals are sprayed widely over large areas, there is the possibility that they may drift on the wind and that small concentrations may build up in the atmosphere at large. They might be hazardous to beneficial insects such as bees, to wild animals and birds which feed in the crop, and to creatures which live within the crop or in the soil beneath it and, thus, indirectly to wildlife which feeds on those creatures. Chemicals which fall on to the soil can be washed down into it by rain and eventually find their way into lakes and rivers and thence into estuaries and harbours, where they might adversely affect fish and other aquatic life. All these are possibilities, the risks of which have to be weighed against the benefits which the pesticides produce. What is clear is that it is as unthinkable for the community to do without modern pesticides as to do without modern medicines. Present-day agriculture in developed countries is dependent on crop protection chemicals as much as it is on the internal combustion engine. No one would want to return to the hunger, malnutrition, disease and discomfort of a primitive existence. However, our goal must be to employ crop protection and pest control chemicals in agriculture, horticulture, public health and amenity and recreational areas as part of the total management of the environment for the long-term benefit and survival of mankind, and to minimise any possible risks arising from their use.

Right from its start in the 1940s the modern pesticides industry has been aware that there were risks and has taken steps to eliminate or minimise them. In the early days, farmers were not made sufficiently aware of the possible dangers of crop protection chemicals and sometimes used them indiscriminately and not too wisely, but the standard of knowledge has risen greatly in recent years. However, there was always the possibility that unscrupulous manufacturers might not give sufficent consideration to the safety of their products or that careless growers might handle and use them in an unsafe way, and it was to prevent situations like these that Governments had to act by means of legislation and regulatory controls to protect the consumer and the community as a whole.

A major responsibility of the Government of any country is to try to ensure that adequate supplies of nutritious and wholesome food are available at reasonable prices to its inhabitants, and it must encourage and promote any technology which can help to achieve this aim. At the same time, it has the responsibility for trying to ensure that application of the technology does not present unacceptable or unreasonable risks to the community or to the quality of our lives. It is inevitable that introduction of any new technology must bring with it the possibility of some new risks, and the aim of Government must be to enact legislation which will minimise these risks without seriously detracting from the benefits. To do this rationally and effectively, necessitates study of cost-benefit relationships of the technology, for the community as a whole. To permit the pros and cons to be weighed up accurately and objectively, it is essential that the study should be quantitative and not qualitative.

Cost-benefit Analysis

It is important to understand clearly what is meant by the term 'cost-benefit'. Cost-benefit analysis of a proposed action requires that all benefits (advantages) and costs (disadvantages) of any kind that could result from that action shall be identified and quantified on the same scale of values, so that they can be directly compared and weighed against each other. Risks of the type to which we have referred are disadvantageous events which may or may not happen, and the cost in a 'cost-benefit' sense is the quantitative effect measured on some scale of values, if such an event does occur, multiplied by the estimated probability that it will occur. Similarly, a benefit in the 'cost-benefit' sense is the quantitative value of the effect of some advantageous event, multiplied by the probability that it will occur.

It is important, also, to distinguish clearly between cost-benefit analysis and investment appraisal. Investment appraisal attempts to measure the effects of a proposed action, such as sale of a new pesticide by a manufacturer or use of a pesticide by a farmer, solely in terms of the cash consequences which it will have for the person who is putting up the money. Unlike cost-benefit analysis which tries to strike a balance over all the effects which the proposed action could have, investment appraisal is concerned merely with the limited region of effects on the investor. The effects which are considered in investment appraisal are sometimes referred to as 'internal' costs and benefits, and those which are only considered in cost-benefit analysis, as 'external' costs and benefits, but this can lead to confusion. The reader should be aware that many studies of pesticide use which have been presented in the literature as cost-benefit analyses are actually nothing more than investment appraisals. A cost-benefit study for the whole community must take into account all possible effects on every section of the community.

Suppose, for example, that a cost-benefit analysis of a proposal to introduce a new pesticide is being considered for the community as a whole (Figure 2.1). First one must list all the effects which this new pesticide might produce, e.g. lower food prices, increase in crop yield and quality, saving of labour, economy in land usage, more spare time for the farmer, contamination of land and destruction of wild flowers, effects on wildlife, etc. In order for the analysis to be meaningful, it is essential to make this list as comprehensive as possible.

Having listed all possible effects in a row down the page, one can then draw vertical columns across the page and put at the head of each the designation of a particular section of the community who might gain or lose, in any way, from any of the effects, e.g. the manufacturer, the merchant, the spraying contractor, the farmer, the farm worker, the consumer of the crops, the conservationists, the general public and the state and federal governments. The gains or losses need not be directly financial but could include such things as increased leisure, more acceptable food, effects on health, variations in the quality of life, affronts to one's beliefs, etc. Once again, it is essential that this list should be as comprehensive as possible. To carry out the cost-benefit analysis, it is then necessary to put into each square a figure in some common scale of values, so that the results of every effect on every affected group is expressed in positive or negative terms. The most convenient and generally used common scale of values is money. It may not prove

Benefits / costs	People affected										
	Manufacturer	Agricultural merchant	Spraying contractor	Farmer	Food processor	Food retailer	Consumer	General public	Conservationist	Government	
Cash gains											
Provisions of employment											
Increase in gross national product–exports											
Safeguarding economic viability of a growing region											
Saving of labour											
Saving of machinery											
Saving of energy resources											
Economy in land use											
Increased leisure											
Increased convenience											
Better supplies of food											
Effects on food costs											
Greater variety of food											
Improved quality of food											
Risks to operators											
Risks of accidents to public											
Health risks of residues											
Risks of contamination of watercourses and rivers											
Effects on soil											
Risks to non-target species											
Risks to wildlife											
Other long-term environmental effects											Grand total
Individual column totals											

FIGURE 2.1   Format for cost-benefit analysis of a pesticide.

possible to fill every square with a cash value but, if many of the most important squares are not filled, the cost-benefit analysis becomes meaningless. If the formidable task of filling each square can be accomplished, the figures in each column can be totalled up to show the net effect on each group, and then these can be added together to give a grand total which represents the benefit or cost to the community as a whole.

It will immediately be apparent that a positive grand total does not necessarily imply that each individual column total will be positive: some may be very positive and some very negative. This means that, although introduction of the new pesticide might benefit the community as a whole, some sections of the community would gain and others would lose. This is a situation which nearly always arises whenever any action is taken which could affect a number of sections of the community and, particularly, when any legislation is enacted. It is the responsibility of Governments to try to mitigate any hardship that is caused to any particular sections of the community, without detracting from the overall benefit to the community as a whole. In other words, they have to try to iron out the positives and negatives in the individual columns of the cost-benefit analysis without altering the grand total. It must always be understood, however, that optimisation of cost-benefit for a whole community does not necessarily imply optimisation for every individual in that community and that, no matter what is attempted, a point will eventually be reached at which an increase in the gain of one individual must result in a corresponding loss for another.

It is also important to understand that the figures in individual squares in a cost-benefit analysis cannot be considered in isolation from one another, and that any action which is taken with the intention of altering the figure in one square will inevitably affect the figures in many, or even all, of the other squares. This is because all the effects and all the various groups of affected people are interdependent. They are part of a total 'system' and, if the system is poked at one point it may bulge at another, often unexpected, place. The mathematical techniques which make it possible to study and predict the behaviour of such 'systems' is called 'systems analysis' and is an integral and essential part of any attempt to optimise a cost-benefit situation.

A system must, of course, have boundaries and it is possible to draw the boundaries very widely (for instance, to consider the effects of pesticides on the total life of a community) or to draw the boundaries very narrowly (for instance, to consider the effects of pesticides on one particular species of wild flower). Both are valid things to do in a cost-benefit study, but it is very necessary to be aware that, if you optimise within the boundaries of a particular system, you are ignoring any effects which your optimisation may have on things or people that are outside your system boundaries. How wide should a system be? It may be argued that the only completely valid system for study is the whole universe and that all other systems are sub-systems. This is certainly true, but it has also been said that only God can optimise—'Not one sparrow shall fall to the ground without your Father's will'—and that anything which men can do must be a sub-optimisation. So, we have to accept these limitations but be constantly aware of the inherent pitfalls in sub-optimisation, which increase as the system is made smaller. Where, then, is it reasonable and practicable to set the system boundaries for a study of the effects of pesticides? The view expressed in this book is that the most useful system to investigate are sets of people who are governed and affected by the same laws, and who have delegated their powers of decision to the same set of legislators. Cost-benefit studies should be made from the point of view of all those people and the immediate environment in which they live. This implies an attempt to optimise cost-benefit for each individual country, since one set of legislators can make laws only with respect to their own populations. This does, however, mean that there would be a possibility that action taken by one country to optimise its own cost-benefit situation for pesticide use might be disadvantageous to another country. There is a great deal of force in the contention that the proper system for study of the effects of pesticides is the whole world—all the people in it and their total environment—but such a study could not be fruitful in the absence of effective methods of world legislation. Nevertheless, it should be regarded as an ideal to be aspired to and worked towards. History reveals clearly the outcome of attempts by individual nations to sub-optimise cost-benefit situations purely from their own viewpoint—such attempts have been a frequent source of conflict and war.

Within the boundaries of any nation, individuals will continually be trying to optimise, as far as they are able, any cost-benefit situation which affects them. This is not to suggest that all individuals are motivated by personal greed since many will be willing to make sacrifices to achieve a result which they consider worth-while either for the community as a whole, or for some section of the community about which they personally care. What they will try to optimise is personal satisfaction. Various groups of individuals with some common interest will also combine to try to optimise a cost-benefit situation from their combined point of view. In developed countries in recent years this has been attempted by lobbying and by pressure groups, and by utilising the mass media to enlist public support. The objective of such tactics is to try to get it accepted that the point of view of the particular group is more important and significant and more deserving of consideration than the points of view of other groups. Such groups may, understandably, pay little regard to optimisation of cost-benefit for the community as a whole. In many cases, the groups are genuinely not aware of the total effects on the community that the action which they propose could have, and they act entirely in good faith in the belief that the matter about which they are concerned could be dealt with in isolation.

It is the responsibility of decision takers in Governments to try to base their decisions on the cost-benefit situation for the community as a whole, and to resist undue pressures from individuals and groups. Thus, in the field of pesticides, it would be wrong for a Government to ban a particular product because it was alleged to have an adverse effect on some form of wildlife, without considering what effect that ban would have on such matters as the economics of farming and the price and availability of foodstuffs. Equally, it would be wrong to allow a farmer to use any pesticide he liked, in any way he liked, without thought for its environmental effects. Politicians are, however, human. Their careers are dependent on the support they receive from the public, so they may have, at times, an understandable temptation to placate those who can apply the most pressure and create the most agitation.

In the field of pesticides, it is vitally important for those technical departments of Government, who have the responsibility of supplying information on which legislators base their decisions, to work towards the aim of building up and presenting a full cost-benefit picture for the whole nation. The extreme difficulty and complexity of doing this and the impossibility of ever achieving it completely should not inhibit attempts to approach it, and the ideal should always be kept clearly in view, even if it cannot be attained. To see that its population is adequately fed is the most fundamental and important responsibility of the Government of any nation, and it is for this reason that it is essential to produce the information on which to achieve a balanced view between the need to utilise crop protection technology to produce more food and the risks which application of this technology might possibly produce for the consumer and for the environment in which he lives. In particular, it is necessary to try to quantify all the benefits and all the costs on a common scale of values so that they can be weighed against each other. The relative importance of each aspect can then be properly assessed.

There are three general problems which are encountered in attempting a cost-benefit analysis. The first is to assess risks accurately—that is, to predict in quantitative terms what the effects of any changes will be on the various groups of people involved. This difficulty arises from the need to make the analysis before the event, rather than after it, if it is to be of any use as a guide to decision and action. This can be met partly, but not entirely, by extrapolating trends from past experience, but this can never give an entirely correct forecast, because the situation is always dynamic, and concurrent changes in external factors inevitably affect the shape of the extrapolated curve. Also, precisely because the situation is dynamic, any cost-benefit analysis will relate only to the time when it was carried out and may become invalid if changes occur.

A second problem is that the value in cash terms of a cost incurred or benefit received today is not the same as the value of some cost incurred or benefit received in ten years' time, particularly in these days of rapid inflation. Money has a time-value and a dollar today is worth considerably more than a dollar promised in several years' time. This problem can be overcome to a considerable extent by discounting cash values at a specified percentage rate for each year they range into the future, so as to bring all values to present-day figures. However, it is not always easy to select the most appropriate discount rate.

The biggest problem is that of converting many of the costs and benefits into cash terms. What, for instance, is the cash value to a particular individual of increased leisure, of blemish-free fruit, or of a wild flower? It can be useful, as a first guide, merely to place a plus or minus in each square of the cost-benefit analysis to indicate a benefit or cost. Then, as a refinement, a simple scoring system can be introduced in which, say, + 3 indicates a very large benefit, + 2 a moderate benefit, + 1 a small benefit, − 1 a small cost, − 2 a moderate cost and − 3 a substantial cost. An examination of the general appearance of the completed table in each case will give a useful impression of the total situation. Scoring systems can be made more and more precise as more information is accumulated. This sequential approach by a series of increasingly refined and detailed scoring systems is a useful and revealing technique.

Despite all these difficulties and despite the complexities, we shall consider, in the following chapters, various matters relevant to cost-benefit analysis of crop protection and pest control from the points of view of the main groups of people who are affected—namely, the pesticide manufacturer, the farmer, the food processor, the consumer and the general public (as trustees of the environment). We shall then discuss the problems that Governments face in drafting legislation and regulations within the framework of which it will be possible to approach the best total results for the nation as a whole. Then will follow a short account of possible alternatives to chemical pesticides for crop protection and pest control. Finally, we shall discuss the matter of public health, which involves a number of factors which are different from those for crop protection, but which is a matter of extreme significance in terms of health and well-being of humans and animals.

3

The Manufacturer's Viewpoint

The modern development of crop protection chemicals came almost entirely from the chemical industry. In this respect it differed from many technologies which evolved from initial discoveries made in the scientific pursuit of knowledge in the universities and other academic institutions. The object of research in industrial companies is not, as in universities, to increase scientific knowledge for its own sake but to find new ways in which the company can increase its business and earn profits. It is relevant to ask why, in the period following World War II, the chemical industry considered that crop protection chemicals might be a source of profit, and why they were successful?

It has already been pointed out that advances made in the science of organic chemistry during the latter half of the nineteenth century led, in the first half of the twentieth century, to commercial exploitation of organic chemicals, first in dyestuffs and then in pharmaceutical products. Simple organic compounds, which were cheap to produce, could be utilised to make plastics, or to treat textiles, or in some other areas of the market for chemicals where cost is a primary consideration. Complex organic compounds are, however, expensive to produce and need much more highly-priced markets if they are to be acceptably profitable. Customers pay for the effects of chemicals and it is obvious that the effect of curing a disease will command a considerably higher price than the effect of producing an ornament or kitchen utensil. For this reason, the chemical industry, particularly those companies engaged in manufacture of dyestuffs, looked to pharmaceuticals as a commercial outlet for the new complex organic chemicals which they could now make, and a number of companies built up thriving businesses on the basis of medicinal products. Before World War II, a number of these companies were exploring the possibilities of using complex organic chemicals in a similar way to control pests and diseases of plants and animals. However, in the 1930s, farming was in a depressed state in the USA and Europe, and it did not seem likely that farmers would be willing to pay the high prices which were needed for such chemicals. Nor was it clear, at this time, that organic chemicals could be found which would be useful in treating horticultural and agricultural pests and diseases. So, it appeared to be a high risk and uncertain undertaking to invest research money in this area. The discovery of the insecticidal properties of DDT by Müller of Geigy in Switzerland, of the organophosphorus insecticides by Schrader of Bayer in Germany and of the phenoxyacetic herbicides by Templeman of Imperial Chemical Industries in Britain demonstrated that useful products could, in fact, be found and that these need not necessarily be very costly. After World War II food prices had risen considerably and farming was becoming a much more profitable occupation, so some dyestuffs and pharmaceuticals companies, inspired by these discoveries, started research on crop protection chemicals as a potentially profitable diversification of their businesses. Furthermore, coal-tar had been for many years the main source of organic compounds which could be used as raw materials to manufacture complex organic chemicals, but by the 1940s the chemical industry was beginning to produce all kinds of new organic starting materials from oil—the so called 'petrochemicals'. Many large manufacturers of primary organic chemicals therefore also came into crop protection as an outlet for their products.

Industrial history shows clearly that commercial success of a new technology does not depend only on its technical merit but demands that economic and social conditions for its acceptance must also be right. This was the case for crop protection chemicals after World War II and particularly for the phenoxyacetic herbicides. These compounds had four advantages which were vital at this period—namely, that they were not exceptionally costly, that they required little skill to use, that they were almost non-poisonous to men and animals, and that they were developed at a time when maximum production of food was essential and farm labour very scarce. The ready acceptance of the phenoxyacetic herbicides by farmers paved the way for the rapid developments of the past thirty years and, without it, these developments would have come very much more slowly. Similarly, DDT was very cheap to produce and was rapidly accepted and widely used by public health authorities for insect control, as a result of which it was applied in a way which helped to create some of the environmental problems, which are associated with this substance nowadays. Acceptance of these cheap chemicals led to the development of markets for the much more complex and costly crop protection chemicals of today. It will be realised how fortuitous it was that the first two major crop protection chemicals—the phenoxyacetic herbicides and DDT—were comparatively cheap to produce. Parathion, which is a highly dangerous pesticide, is still widely used because it is cheap and effective.

As the development of crop protection chemicals was not based on any firm scientific background, the search for them has been, and still is, empirical. Chemical compounds are made at random or according to the 'hunches' of the chemists concerned, and are tested to see if they have any biological activity which might be the basis of commercial utility. The scientific approach of studying in detail the plant and the pest or disease and then trying to design a chemical compound which would effect control has been considered, but the problem is complex and expensive to investigate and results have, so far, been unrewarding. In fact, the search for new pesticides nowadays is not entirely empirical, because a great deal of knowledge has been accumulated over the past thirty years which can be used to guide research to home in on the most likely types of active compounds. Nevertheless, chance and luck still play a great part.

The procedure by which new crop protection chemicals are developed is to set up in the glasshouse 'screening' tests which are designed to cover a wide range of plant/pest/disease situations and, hopefully, to reveal any biological activity of possible commercial utility in any of the compounds submitted to these tests. Candidate compounds which show promising activity are taken on to secondary and tertiary evaluation procedures in the glasshouse which examine the observed activity in greater detail, with special regard to activity and selectivity at various application rates. The term 'screening' is appropriate as it is similar to putting something through a series of wire screens of smaller mesh. Surviving compounds then go out to initial field trials. A major problem with this type of development is that the results obtained in the glasshouse are often not reproduced in the field. Field trials are, by nature, seasonal and, if a satisfactory result is not obtained one year, one has to wait another twelve months before the trials can be repeated. In subsequent years, field trials are carried out on an ever-widening scale, in different countries, with different climates and different types of soil, in order to identify the key commercial uses and their likely problems and limitations. Concurrently with these field trials, extensive toxicological and environmental studies are set in motion as soon as the first field trials are contemplated. Very detailed studies have to be made of the effects of the chemicals on humans and animals, on the possibility of residues in the crop at harvest, of effects on all the microfauna and microflora of the soil and of possible dangers, either directly or indirectly, to wildlife of any kind. During the same period, intensive biochemical studies are made to try to understand the way in which the compound acts, and formulation studies to find the most satisfactory way in which the compound can be presented for use. Research also has to be undertaken to discover and develop economically viable processes for manufacture of the chemical, and equipment has to be set up to make pilot quantities. The complex interrelationships of all aspects of development are shown in Figure 3.1. When all this work has been done, application for permission to sell can be made to Government registration authorities, who may often require yet more tests and investigations before they are satisfied.

Development of a new crop protection chemical is, therefore, a very long and expensive business. In Table 3.1 some information is shown

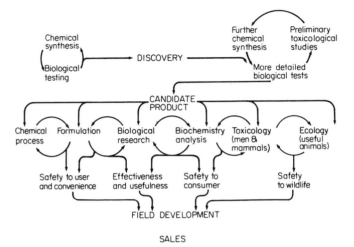

FIGURE 3.1   The evolution of an agricultural chemical. (By courtesy of Plant Protection Limited, England.)

from a survey made in 1971 of thirty-three major pesticide companies in the USA. This shows, for 1970, an average of 7430 compounds tested for every one which was eventually developed commercially, a total discovery and development cost for each product of about $5·5 million and a total time from discovery to registration of about 6·5 years. All these figures have been increasing steadily from year to year as shown in Table 3.2. One major reason, apart from the rising number of compounds which have to be screened to get a commerical product, why the cost and time of development of a new pesticide are both increasing is the very great expansion of the amount of toxicological, residue and environmental data that has to be provided for the registration authorities. This is illustrated in Table 3.3.

Table 3.4 shows some estimates for 1970 which take into account the

TABLE 3.1
Research profile of US pesticide industry. (Ernst & Ernst Trade Association Department, May 1971)

	1967	1970
Total sales	$639 m	$722 m
R & D on new products	$52·3 m	$69·9 m
Percentage	8·2	9·7
No. of compounds screened per marketed product	5481	7430
Cost of development of each marketed product	$3·4 m	$5·5 m
Time from discovery to marketing	60 months	77 months

TABLE 3.2
Increase in development costs of a pesticide

Year of estimate	1956	1964	1967	1969	1970	1972
Cost of development ($ millions)	1·2	2·9	3·4	4·1	5·5	10·0
Number of compounds screened per marketed product	1800	3600	5500	5040	8000	10 000

TABLE 3.3
Minimum requirements for world-wide registration

	1950	1960	1970
Toxicology	Acute toxicity 30–90 day rat-feeding	Acute toxicity 90 day rat-feeding 90 day dog-feeding 2 year rat-feeding 1 year dog-feeding	Acute toxicity 90 day rat-feeding 90 day dog-feeding 2 year rat-feeding 2 year dog-feeding Reproduction 3 rat generations Teratogenesis in rodents Toxicity to fish Toxicity to shellfish Toxicity to birds
Metabolism	None	Rat	Rat and dog Plant
Residues	Food crops 1 ppm	Food crops 0·1 ppm Meat 0·1 ppm Milk 0·1 ppm	Food crops 0·01 ppm Meat 0·1 ppm Milk 0·005 ppm
Ecology	None	None	Environmental stability Environmental movement Environmental accumulation Total effects on all non-target species

TABLE 3.4
Estimate of development costs of a pesticide

Year		No. of compounds	Cost per compound	Total cost
			$	$'000's
1	Synthesis	8000	140	1640
	Screening		65	
	Survival rate 1:100			
2	Glasshouse trials	80	5000	400
	Initial field trials			
	Survival rate 1:5			
3	Field trials	16	5000	112
	Initial toxicology		2000	
	Survival rate 1:4			
4	Field evaluation		50 000	
	Toxicology	4	20000	480
	Formulation and process		50 000	
	Survival rate 1:2			
5	World wide evaluation		500 000	
6	Toxicology, environment, ecology		350 000	
	Formulation and process	2	200 000	2850
	Production		200 000	
	Registration and patent		175 000	
	Survival rate 1:2			
			Total cost	5482

fact that each successful product has to pay for the 7429 which are tested but eventually fail; and some of these have been discarded, not at the screening stage when little money has been spent on them, but only after several years of expensive trials and toxicological studies. The rough time scale shown alongside the figures is for an idealised case in which everything has gone well. Not only must compounds be evaluated over a range of seasons, because no two seasons are alike, but the results of one season's trials are also often inconclusive and contradictory, so that they have to be repeated, which sets the whole project back, and makes it much more costly.

The capital which has to be invested to produce a pesticide is not small, since most new pesticides are fairly complex organic chemicals requiring several stages in their production and the use of complicated and expensive equipment. Also, because they tend to use costly raw materials and because the value of the final product is high, large amounts of working capital must also be provided. A typical figure for total capital required for manufacture of a pesticide in 1970 was $3000 per tonne per year, compared to about $300 per tonne per year for, say, a plant for manufacture of polythene.

A manufacturer developing a new pesticide, therefore, has to put in large amounts of capital and revenue and wait a long time for the rewards. This is illustrated in Figure 3.2 by a discounted cash flow, which shows the cumulative difference between outgoings and incomings of money

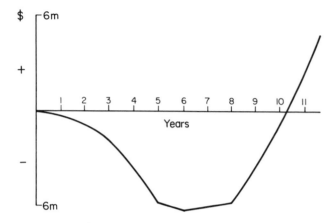

FIGURE 3.2   Discounted cash flow of successful pesticide.

over the whole project life, adjusted to take account of the fact that a dollar in 10 years is not worth the same as a dollar today. This is the cash flow for a winner, the development of which has gone smoothly. First, there is the comparatively low cost of the initial years of discovery. The cash flow includes only the costs of the particular compound, since the costs of abortive work on the 7429 compounds which failed must be met from current profits on established products and cannot be offset against profits which might or might not be made in ten years' time. After the initial period, the cash flow drops sharply when capital is invested in pilot plant and when the prolonged and expensive toxicological, ecological and environmental studies are undertaken to ensure safety and freedom from hazard. The cash flow dips even more sharply when larger production plant is built four to five years after the discovery. Then come the first years of sales when costs are just being covered, followed, if the product is successful, by a sharp rise to the break-even point at which all money invested has been recovered. After this comes a period when a level of profits is reached which justifies being in the business at all and which pays the high costs of ongoing research to discover other useful products and to develop the next generation of pesticides and pay for all the work done on compounds which are tested but discarded. The pesticides industry is, therefore, a 'research-intensive' business and research and development costs average about 9·7% of total sales turnover. The period of high profits may not last long because better competitive products may appear, resistance problems may develop, techniques of farming may change and, in any event, the patents—which will have been taken out when the product was discovered ten years earlier—will expire and allow other manufacturers who have had to bear none of the financial burden to come in and force down prices.

Three factors make the years of first sales unprofitable. One is that production of small tonnages of complex chemicals is relatively very expensive because labour, maintenance and service costs per unit of output fall considerably as the size of a chemical plant is increased. Essentially, this is because one man can just as easily look after a big reaction vessel as a small one. A typical example is shown in Figure 3.3.

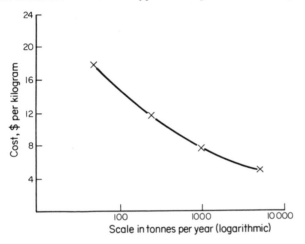

FIGURE 3.3   Variation of cost with scale of manufacture.

The second factor which increases costs in these years is that the full costs of production depend on the extent to which the capacity of a chemical plant is utilised. A typical example is shown in Figure 3.4. It is no use building a plant which will satisfy only this year's demands; the size of the plant has to be such that it can satisfy expected demands for several years ahead. This means that, during the period of build-up of sales, the plant will only be partially occupied for several years and this increases the costs of the low tonnages needed in the first years of sales and means that, during these years, the manufacturer will be lucky to break even.

The third factor which makes the initial sales period unprofitable is the very high costs of promoting and advertising a new product. There is often a natural conservatism, especially amongst farmers, which makes them very wary of anything new and it needs extensive demonstrations, personal visits by representatives, direct mailing and articles in magazines and journals to convince potential customers that the product is worth buying.

How then does the manufacturer view cost-benefit analysis as it

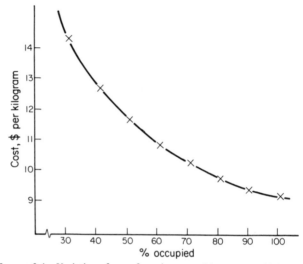

FIGURE 3.4   Variation of manufacturing cost with extent to which plant is occupied.

applies to him? He looks at it mainly on an investment appraisal basis and wants to know the probability of incurring the estimated expenditure during the development period and the probability of achieving the estimated profits eventually. What is quite clear is that the costs are a lot more certain and a lot nearer in time than the benefits.

Consider a manufacturer standing at a point on the cash flow in Figure 3.2 one or two years after discovery of a new pesticide and trying to estimate what the shape of the cash flow will be and what range of variations may occur. The available information on selling prices, rates of growth of sales, application rates, production costs, etc., will be very uncertain at this stage, and many unexpected things can happen in ten years. Mathematical techniques are available for dealing with such high risk situations and these quantify the gamble; but they do not prevent it being a gamble, nor do they prevent wrong decisions being taken and large amounts of money being lost. The average cost of development of a new pesticide was $5·5 million in 1970, but the chance that it would exceed sales of $5 million per year was about 10 to 1 against. Figure 3.5 shows the sales in Western Europe in 1970 of all pesticides introduced between 1960 and 1970; 12 exceeded $2·5 million while 36 were less, and the picture for other parts of the world is similar.

To put large amounts of money at high risk for long periods in the hope of eventual profit is acceptable only to the largest manufacturers who

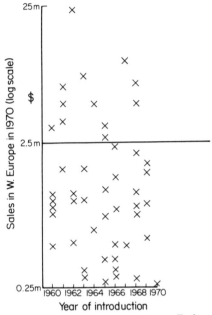

FIGURE 3.5   Sales of pesticides in Western Europe. Each cross represents one product. (By courtesy of Plant Protection Limited, England.)

have a great liquidity of assets and a diversity of interests to buffer them against misfortune. This is why the number of firms engaged in discovery and development of new pesticides is comparatively small, and is getting smaller, and why these are the large diversified organisations. Such reputable companies do not consider cost-benefit entirely in terms of investment appraisal but accept a very large degree of social responsibility for their operations to protect their good names and the reputation of their countries. It is doubtful if any industry spends a greater percentage of its profits than the pesticides industry merely to try to ensure greater safety of their products to the consumer and to the environment in which he lives. This is not entirely altruistic, since no company is going to invest the large amounts of money which are needed for development of a new pesticide if there is a risk that, at a late stage when much money has been spent, the pesticide has to be discarded because of some unforeseen hazard to the consumer or to the environment. The manufacturer will therefore take steps to investigate these problems thoroughly at the earliest stage possible.

However, such companies have a legal responsibility to their shareholders to invest their money wisely; and if discovery and development of new pesticides cease to be acceptable risks, they might well decide to discontinue these activities and invest their money in other interests—pharmaceuticals, plastics, textiles, etc. They have no obligation to go on making pesticides and nobody should expect them to do so if the cost-benefit analysis is obviously unfavourable for them.

There is a considerable possibility that this situation may come about. A large number of pesticides have been produced during the past thirty years and many of these have now achieved large tonnages and have become patent-free so that they can be made comparatively cheaply because of scale of manufacture and competitive production. Many other pesticides will move into this category during the next ten years. There are, therefore, reasonably satisfactory answers available to a large number of crop protection problems. The available markets for pesticides will, because of those trends, become increasingly fragmented in the future. There will be smaller market opportunities and lower tonnages for new pesticides. However, because of the high costs, discovery and development of a new pesticide are acceptable undertakings for an industrial organisation only if the product can command a large market in a major crop. This is partly because costs of production at low tonnages are high and partly because the costs of searching for new biologically active compounds and the costs of toxicological, residue and environmental studies are generally independent of the size of the ultimate market. From an investment point of view, it is not justifiable for a manufacturer to deploy resources or invest money to discover solutions to crop protection problems where the market opportunity is small, or even to develop products when a likely solution emerges from primary screening and evaluation procedures—unless the selling price can be high enough to produce an acceptable cash flow or unless his cost-benefit position can be mitigated in some other way. This is already the situation with regard to pesticides for minor crops, and many products which could be useful or safer for these are being left on the shelves of the research departments. Many of the crop protection problems in the future will be economically similar. For instance, a problem pest in one small part of the country in a major crop grown over a wide area is essentially a minor crop problem. Yet, for a particular region which has that problem, it may be a matter of economic survival and the continued well-being of the inhabitants to find a solution.

Use of pesticides to solve such problems must be subject to the constraints of cost-benefit from the viewpoints both of the farmer and of the consumer. The price which a consumer will pay for a foodstuff sets an upper limit to the price which a farmer can afford to pay for a pesticide. The situation may increasingly arise that a pesticide which, on a cost-benefit analysis, would be very beneficial to the community may, on an investment appraisal basis, be quite uneconomical for a manufacturer to produce. When we come to consider cost-benefit from the point of view of the community as a whole we shall be forced to the inevitable conclusion that, if the public wants to go on enjoying the benefits of pesticides without hazards to themselves or to their environment, they will have to be prepared to pay more for those benefits. This could come about either by direct increases in food prices or indirectly through taxation by Government subsidy of pesticide manufacture or by Government organisations such as the United States Department of Agriculture taking a great deal of the financial burden of toxicological, residue and environmental investigations off the shoulders of the manufacturers. As the requirements of environmental safety get more and more stringent, these investigations become more and more expensive and prolonged, and the combined effects of these two factors on cash flow makes achievement of a satisfactory return on investment more and more difficult.

The list which follows gives some of the environmental aspects on which satisfactory and acceptable data must be provided to the Environmental Protection Agency before a new pesticide can be registered for commercial sales in the USA. This list does not necessarily cover all questions relating to the environment which may be asked and it is, of course, only a part of the total information which is required by registration authorities which includes extensive data on toxicology, residues, mode of action, etc. It will be appreciated that the collection of all this information necessitates a very formidable amount of work. This is especially true when it is appreciated that all the work has to be done not only with the pesticide itself but with every chemical compound into which this pesticide can be broken down in the soil or within the plant—and there may be as many as twenty of these.

The questions to be considered in evaluating the impact of a pesticide and its degradation products on non-target organisms are:

1. What is the mode of action in the target species, taxonomically closely related non-target species, and/or taxonomically unrelated non-target species? Is the mode of action species specific (i.e. affecting a metabolic process characteristic to the target species only) or is it operative through metabolic processes common to a wide variety of taxa?

2. How broad is the spectrum of biocidal activity to non-target species closely related to the target species or to other non-target species widely separated taxonomically?

3. Will the candidate pesticide cause significant habitat alterations under the conditions of expected use?
4. What behavioural effects are likely to occur with non-target species present in treated areas and will these behavioural effects affect the species' survival capabilities?
5. What is the persistence of the pesticide or its degradation products in the environment and what relationship does this have to its mobility in air, soil, water and biological organisms?
6. What is the likelihood of bioaccumulation and biomagnification to levels producing significant effects in non-target species?
7. What metabolic processes are most likely to be affected and will these processes cause diseases, lowered disease resistance, selection for or against development of resistance to the pesticide, changes in growth patterns, etc.?
8. What is the specific relationship of the candidate pesticide to reproductive potential and success?

In summary, the questions which will concern a manufacturer in any comparative cost-benefit study of manufacture of various pesticides are as follows:

1. Will production and sale of this pesticide produce an acceptable discounted cash flow? Will the rate of growth of sales and the selling prices which can be achieved, and the ultimate size of the market adequately recover the research, development and marketing costs and provide an acceptable return on capital invested before the patents expire?
2. What is the competitive position of the product? How long can the market be retained?
3. Are any toxicological or environmental problems likely to arise which could cause the pesticide to be withdrawn or harm the reputation of the Company?
4. Has the Company a better investment for its money than pesticides?

4

The Farmer's Viewpoint

Development of farming has taken place over the past 6000 years. During that time farmers have discovered by experience techniques by which the effects of some pests and diseases on various crops can often be minimised. As there were few known methods by which pests and diseases could be attacked directly, these techniques have been mainly cultivational and agronomic. Thus, rotation of crops was devised to prevent build up of large populations of a particular pest in the soil. Farmers learnt to adjust depth of sowing of seeds, date of sowing, supply of nutrients and date of harvest in such a way as to lessen attacks. Ploughing provided a clean tilth for sowing and reduced competition from weeds. Those weeds which did appear were kept under control as far as possible by manual weeding and hoeing. Farmers gradually came to recognise the importance of hygiene and of avoiding waste heaps or large patches of weeds which might serve as alternative or overwintering sites for pests and diseases. Over the years, farmers experimented with cross-breeding of animals and plants, mainly in order to obtain better-yielding varieties, but they discovered incidentally that some of the new varieties had increased resistance to some pests and diseases. This was the foundation of the modern science of plant breeding.

The development of modern pesticides has given the farmer a whole new armoury in his fight against pests and diseases and has made possible the introduction of new techniques of cultivation which could not be used previously, such as successive monoculture and very high planting densities. Nevertheless, pesticides have not made the older methods useless but have provided new alternatives for the farmer in any cropping situation. He will continue to use traditional techniques if they are effective since pesticides cost money and there is no justification for using them if they do not produce adequate benefits, that is, unless the cost-benefit situation is favourable. There was, in the earlier years of modern pesticides, some tendency amongst some ill-informed farmers to use pesticides just because they were new and available, without considering sufficiently whether there was a genuine need to do so. This tendency may have been to some extent increased by over-enthusiastic sales promotion by the manufacturers. Nowadays, mainly as a result of educational and information services produced by the advisory and extension branches of the Government agricultural departments, pesticides are, by and large, being used by farmers in a sensible and economical manner. In the rest of this chapter we shall discuss the various considerations which influence a farmer in his decisions on crop protection and pest control methods.

The farmer's immediate view of crop protection is in investment appraisal terms, that is, an assessment of the extra cash he may receive for his crop by preventing or controlling weeds, fungi and insects compared with the cost of the crop protection treatments needed to do this adequately. To be able to make such an assessment accurately he would need to know, for each crop:

1. The probability of attack by each pest and disease during that particular growing season.
2. The probable severity of attacks if they do come, that is, by how much they are likely to reduce his crop yields or by how much they are likely to impair the quality of his crop.
3. The probabily of having his crop yields diminished by other causes such as drought, severe storms, etc.
4. The probable range of prices at which he will be able to sell his crop according to quality.
5. The costs of the various crop protection treatments (including the costs of application) available to him and their probable effectiveness.
6. The minimum amount of each crop protection chemical he would have to use to obtain adequate control of each pest or disease.

It is obvious that it would be quite impossible for a busy farmer to collect all this information for every possible pest and disease situation in every crop he intends to grow, and, even if he could, it would be beyond his ability to carry out in each case the very complex mathematical analysis which would be required to arrive at estimates of the probable cash gains or losses which might be produced by each course of action which is open to him. Such figures would, in any case, not indicate gains or losses with certainty but merely show the size of the gamble he would be taking. In practice, farmers do not behave this way, so how do they reach decisions on crop protection?

In developed countries in recent years farms have tended to become larger and more specialised, and the small farmer who kept various types of animals and who grew a wide assortment of crops is tending to disappear. The facility with which modern crop protection methods permit monoculture has turned many farmers into growers mainly of one type of crop. Thus there are wheat farmers, corn farmers, soya bean farmers, beet farmers, cotton farmers, vegetable farmers, etc. By and large such specialist farmers have a fair idea from their own experience, or from that of their neighbours, of the types of pests and diseases which are most likely to occur in their crop in their particular locality. They will also know that there are a limited number of crop protection methods which have proved effective in these various situations in the past, so, in practice, their choice of action is restricted.

Farmers will think subjectively about points 1 to 6 above using their experience and intuition. A great deal of relevant information is available to them from Government advisory and extension services, from growers' associations, from food processors and from technical representatives of manufacturers. Thus, in connection with points 1 and 2 (the possibility of attack by pests and diseases and the probable effects of any attack on crop yields and quality), the farmer will often have the benefit of advice from forecasting services. Unfortunately, although such services certainly save growers considerable amounts of money, they are far from precise and most people who attempt to develop them tend to have much less faith in them than the people who use them. There is such a large area of doubt that the forecaster always tends to play safe and advise the use of pesticides in any circumstances in which he is not sure.

Accurate forecasting would require a very considerable insight into the population dynamics not only of the pest or disease but also of any predators. This is a very difficult and complex subject and it is doubtful if all the predators of any one pest have even been identified, and the delicate balances and interactions which determine population numbers from day to day are almost an unknown quantity. Before any sort of forecasting can be attempted, accurate measurements have to be made over several years in various regions of the numbers of the pest or amount of the disease, the amount of damage caused and the correlation of this with crop yield and quality. This is a time-consuming task and local factors can influence the results. For example, the vagaries of local weather can cause very rapid changes in infestations of certain pests and diseases, for example, of aphids.

In connection with point 3 (the risk of crop losses by other causes), the farmer's decision on crop protection will be greatly influenced by his attitude to risk. For instance, in subsistence agriculture in developing countries where the level of total investment in the crop is small, there is a very narrow range of possibilities for profit or loss so that the effect that crop protection can have on the total financial situation may be small, although it can, of course, represent the difference between survival and starvation for peasant families. In more technically advanced agriculture, in which the investment of resources is very much higher, there are possibilities of much larger profits, but the extent of losses from poor yields is far greater, so that crop protection can have a much larger total financial effect.

In connection with point 4 (the probable range of prices at which he will be able to sell the crop according to quality), the farmer is in no doubt in those cases where he is growing under contract, and this practice is increasing. Also, in the case of cereals, most Governments operate guaranteed prices schemes so that the price which the farmer will get can be predicted fairly accurately from year to year. With vegetable and fruit crops sold on the open market there can be very wide fluctuation in prices which make the economic assessment of the value of crop protection very difficult for the farmer. An example has been given by Coaker and Wheatley relating to production of brussels sprouts in the UK in 1968–69. They calculated total growing costs for a number of farms and arrived at an average cost of $90 ± 7 per tonne of which about $13 were for crop protection. They then recorded the average weekly prices for brussels sprouts in the 1968–69 season and found them to be distributed as shown in Figure 4.1. The average return was about $126 per tonne.

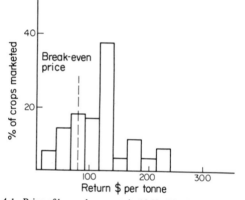

FIGURE 4.1 Price of brussels sprouts in 1968–9 in the UK. (By courtesy of G. A. Wheately and T. H. Coaker, National Vegetable Research Station, England.)

However, it will be seen that about 30% of the returns were below the break-even price of $90 per tonne so that, in these cases, increased crop protection would only have increased financial loss, and the data recorded does not take into account those crops which were ploughed in and became a total loss. In circumstances such as these, economic assessment of crop protection becomes completely meaningless and the farmer can only rely on experience. Any moves by Government to make agricultural production more guaranteed with regard to yield, uniformity, quality and price of the product and to timing of the markets may help to increase the possibilities of increasing economy and safety in pesticide use. The economic constraints are themselves an effective deterrent to excessive and prodigal use of pesticides.

In connection with point 5 (the costs and effectiveness of the various crop protection treatments available), the farmer is subject to some confusion from the vast number of proprietary brands of pesticides all with different brand names. For example, in the USA there are approximately 400 approved pesticides but over 60 000 registered labels. The Standards Institutes of the various countries together with the International Standards Organisation are doing valuable work in attempting to get internationally agreed common names for all pesticides which will appear on the labels so that the farmer can be certain of what he is getting. Also, the regulatory authorities of most developed countries exercise control of the ways in which products are labelled and aim for clarity and precision.

The problem of determining the effectiveness of a crop protection treatment is a difficult one since the farmer normally does not run a control experiment by leaving a portion of his crop untreated, so he is never certain exactly what he has achieved. Effectiveness is usually determined by manufacturers by extensive evaluation trials using various application rates and various methods of application on randomised untreated and treated plots within a crop. These trials are carried out in many different regions throughout the world so that the effects of climate, soil types, varying plant/pest relationships, etc., can be assessed. By taking the results of such trials in conjunction with statistical data on the observed incidence and level of infestation in particular regions it is possible to approach values for the average gains in crop yield which a particular crop protection treatment may be expected to produce. When such treatments are introduced into routine use, the benefits they are producing can be examined by carrying out surveys at frequent intervals. The results of such surveys are, however, difficult to interpret unequivocally since factors other than crop protection may have significant effects.

In connection with point 6 (the minimum amount of pesticide needed to obtain adequate control of a pest or disease), the special consideration is how much damage is acceptable. Such damage can be defined in terms of (a) the percentage of crop units damaged and (b) the amount of damage on each affected unit. Figure 4.2 shows how the percentage of damaged units in a typical treated crop is related to the level of pest attacks and the efficiency of the control method used. If 10% damaged produce can be accepted, a control measure which reduces the population by 90% would be sufficient to protect a crop from an attack of an intensity which would damage about 65% of untreated produce. If, however, no more than 1% damaged produce could be accepted, the population of the pest would have to be reduced by 99%. Such figures can be interpreted in terms of the application rate of a particular pesticide needed to produce the desired reduction in population by use of the data obtained from manufacturers' evaluation trials.

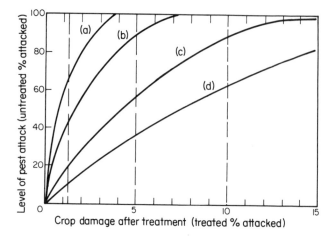

FIGURE 4.2   Relationship between level of pest attack, efficiency of pest control and crop damage after treatment. Efficiency of control (a) 99%, (b) 98%, (c) 95% and (d) 90%. (By courtesy of G. A. Wheately and T. H. Coaker, National Vegetable Research Station, England.)

Whenever a farmer contemplates growing a particular crop in a particular locality, he has three choices open to him:

1. To use no crop protection methods at all.
2. To use preventive crop protection methods, that is, treatments which guard against the possibility of attack by a specific pest or disease. For example, he could purchase seed treated with chemicals to protect the crop against certain fungal diseases during its whole growing period, or he could add a herbicide to the soil when he sows the seed to destroy any weeds which might germinate along with the crop.
3. To use remedial crop protection methods only when his crop is actually attacked by a pest or disease.

If the farmer decides on preventive action he is, in effect, taking out an insurance policy to guard against the risk of loss, just as a householder may take out an insurance against the possibility of loss by fire. He is, like the householder, spending a certain amount of money now to guard against a financial loss which may or may not occur in the future, and which he, in fact, hopes will not happen. To make this decision he has to assess from his own experience, or from the experience of others, of growing that particular crop in that particular location what the chances are that it will suffer from weeds, fungi or insects and, if it does, how severe the damage might be. In real situations the matter is never clear-cut and the most that the farmer will generally be able to do is to estimate what is the worst loss that might occur. If there is a real risk of total or almost total loss of crop this will strongly influence his decision on whether or not to apply preventive crop protection methods. His attitude will be the same as that of a householder insuring against fire in that he wants to be adequately compensated for total loss and is not too much concerned about minor damage. His decision will be further influenced by whether effective remedial measures against the pest or disease which he is considering are available, and how much these cost compared with preventive treatments.

If the preventive crop protection treatment is not aimed at guarding against a pest or disease which might produce total loss but merely at increasing crop yield or reducing crop losses by a certain amount, then the decision becomes more complex. The farmer will need to estimate the price at which he will be able to sell the crop so that he can translate into cash terms the increase in yield which he expects to get as a result of the crop protection treatment. He will also have to consider the possibility of crop losses from other sources. It is of little benefit to safeguard a cereal crop by preventive pesticides against, say, a 20% loss in yield from certain pests and diseases if the whole crop is made unharvestable by autumn gales or is ravaged by some pest or disease which he has not guarded against. A householder would, for instance, hesitate to take out a large fire insurance on a building which was in danger of collapse from structural defects.

To rely entirely on remedial crop protection methods may save the farmer money in that they may never, in fact, become necessary. However, the farmer needs to be sure that he can detect the advent of a pest or disease and take remedial action sufficiently quickly to avoid any significant damage or loss. This could be highly inconvenient to him if it becomes necessary at a time when he has many other things to do on the farm, especially in the current situation of labour shortage. Considerations of labour-saving and convenience may therefore bias a farmer in favour of preventive measures even though it may involve spending cash which might be unnecessary. This implies that he has moved away from thinking in mere investment appraisal terms to a consideration of cost-benefit. From the farmer's point of view there is a lot to be said for a definite programme of preventive crop protection since this can be budgeted for accurately and planned systematically. Ideally, the farmer would like to cultivate the soil, sow the seed, apply the preventive crop protection measures, close the gate and not go into the field again until harvest time.

The overall problem of agriculture is essentially one of increasing productivity per man employed and per unit of land used, at the same time maintaining or increasing fertility in its broadest sense. Labour in all developed countries has been steadily drifting away from the land and it is unlikely that this trend will be reversed. It is likely increasingly to be one of the limiting factors in agricultural production. There is no doubt about the saving in labour which pesticides, particularly herbicides, can bring. For example, in the USA it currently takes about 50 man hours to grow and harvest a hectare of cotton compared with 300 man hours twenty years ago. However, the effect of labour saving on cost-benefit of crop protection treatments on the individual farm can vary from nil, in the case where there is no alternative use for the labour saved, to savings far in excess of the cash value of hours saved in the case where reduction in labour requirements at a critical time can allow advantageous changes in the overall organisation and cropping programme of the farm.

The net effect of a successful crop protection programme is to release the extra land which would have been required to produce the same amount of crops in the absence of control measures. This is a benefit both to the farmer and to the community, and the farmer will take the extra crops which he can produce on this freed land into account in his decisions on crop protection. Alternatively, he may be influenced by the rapidly increasing costs of agricultural land and the high interest rates in recent years which put a premium on intensive cultivation and high yields from small acreages.

It is convenient to discuss at this point two commonly raised objections to modern crop protection methods: (1) that any risks associated with the use of pesticides could be avoided simply by bringing into cultivation extra land to produce the total amount of crops required, and (2) that the use of pesticides will simply result in overproduction of foodstuffs in developed countries with a consequent fall in prices to the farmer who may even be unable to sell his produce.

The first objection could be valid only if the effect of a particular crop protection treatment was merely to increase yield rather than to protect the crop against total loss or total unacceptable damage. Even then, it is based on the assumption that extra arable land is obtainable and that extra labour would be available to cultivate it. Each year in the UK about 20 000 workers leave the land, and urban and industrial developments take over 20 000 hectares of agricultural land. In developed countries, particularly in Europe, there is a chronic shortage of agricultural land. It has been suggested that, even with the optimistic assumption that all possible land is utilised, there will be a desperate land shortage before the year 2000 if *per capita* land requirements and population growth remain as they are today.

The tendency towards overproduction of certain foodstuffs in developed countries has been the cause of some national and inter-national action. Recent examples are the recommendations of the

Canadian Government Working Party to cut the acreage of Prairie wheat by 33% and of the European Economic Community Commission to withdraw 5 million hectares of land from agricultural production in the 1970s. In the USA in the peak year of 1962 26 million hectares were retired, although this trend has recently been reversed. In the face of a decrease in land cropped, the individual farmer must aim at high output from his remaining land if he wishes not only to maintain his income but also to increase it in line with the rest of the community. He will tend, in these circumstances, to adopt more intensive crop protection methods and this is reflected in the fact that, in the USA, despite the reduction in cultivated land, use of pesticides has increased by 15% per year.

Another important aspect which the farmer will consider in his cost-benefit thinking is the possible environmental effects of his crop protection operations. Most farmers care for wildlife and for the countryside which is both their home and their source of livelihood, and would not wish to do anything which is demonstrably damaging to it. Educational programmes can make them aware of possible hazards in what they do and avoid mistakes made in ignorance, such as promoting soil erosion by overcultivation. Farmers are very much aware nowadays of their social responsibilities to use pesticides in a way which minimises possible environmental harm—for example, by run-off from farmland into watercourses, by uncontrolled spray drift, etc. The public have been made very conscious of the alleged risks of pesticides and the farmer has no wish to antagonise them. Also, the farmer has a social responsibility towards the region in which he lives to control the spread of any pest or disease and not to allow anything on his farm to become a possible focus of infection. For instance, badly weed-infested fields may create a major weed problem in subsequent years in the whole area. Crop protection may, in many ways, be regarded as one aspect of farm hygiene.

A final consideration which will be in the farmer's mind is the safety of his workers who have to apply the pesticide. This will also be a matter of major concern to any spraying contractors which the farmer may employ. In the UK a number of fatal accidents during spraying with the highly toxic dinitrocresol herbicides just after World War II led to the introduction of legal safety requirements, culminating in the Agriculture (Poisonous Substances) Act of 1952, which was designed to protect agricultural workers by ensuring that, wherever necessary, they are provided with effective protective clothing and safety equipment. Similar legislation was introduced in most other developed countries. As a result, since that time, the total number of cases of injury to agricultural operatives from pesticides has been very small indeed. No deaths have been reported in the UK since the legislation of 1952.

From the foregoing discussions it will be apparent that a farmer makes cost-benefit decisions on crop protection treatments subjectively taking into account all kinds of personal considerations and views. He will consider crop protection not in isolation but in relation to the total operation of his farm, and it is for this reason that cost-benefit analysis of pesticide use in relation just to one crop can be very misleading. It is necessary to regard pesticides as merely one input into the whole farming operation and to adopt the system analysis approach. People who carry out research on this type of problem try to construct mathematical models of the farm situation so that they can study the interplay and interrelationship of all the factors which concern the running of a farm. However, as we have said, farmers do this subjectively and consequently each individual farmer arrives at an answer which represents a near-optimum from his point of view.

In practice, this will generally lead a farmer to adopt a programme of proved preventive crop protection treatments to guard against the pests and diseases which he knows from his experience are most likely to attack his crop and to rely on remedial treatments for any unusual or unexpected pest or disease which may occur. He will tend to apply a particular pesticide at a particular time simply because his experience tells him to do so. So, by and large, accepted methods of crop protection become established for each crop and the farmer's decision will relate to a very small range of pesticides. This is illustrated by a study made in the USA in 1964 in which 10 000 farmers were individually interviewed and from which it emerged that two herbicides, six insecticides and one fungicide comprised by far the major part of all crop protection chemicals used.

How does a farmer acquire his knowledge of crop protection methods? He may consult with his neighbours and draw upon their opinions and experience. In this regard, the smaller farmers will tend to copy those who are recognised as the local leaders of the farming community. He may get advice from farming journals or from organisations such as the National Farmers' Union in the UK. He may receive help from growers' associations or from associations related to a particular agriculturally-based industry and from the food processors who buy his produce. Thus, for example, the Sugar Beet Corporation in the UK gives extensive help to farmers in all aspects of cultivation and care of their crops. The farmer may seek the opinion of the merchant with whom he normally deals and he will be approached by the technical sales representatives of many pesticide companies and will tend to listen more sympathetically to those with whom he has had satisfactory dealings in the past. He may consult literature issued by Government, or official publications such as the *Weed Control Handbook* and the *Insecticide and Fungicide Handbook*, published by the British Crop Protection Council in the UK. In particular, many farmers will rely heavily on the advice given to them by local representatives of the Government agricultural advisory and extension services. Finally, if he is growing a crop under contract for a food processor it is likely, as we shall see later, that a schedule of crop protection treatments will be specified for him so that the decision is taken out of his hands.

How do new pesticides get into circulation? A farmer may be prepared to try a new product of a company whose existing products have given him good results. He may get advice on new treatments from any of the sources listed above. In general, the most go-ahead farmers in an area will try new products as they come into the market and, if they are successful, their utility in a particular crop in a particular location will become established as part of the general farming experience in that region. Nowadays many Governments operate educational and training schemes for farmers.

It is obvious that widespread use of preventive crop protection treatments results in large amounts of pesticides being applied when, in the event, they are not actually needed, and that this increases the possibility of some risk to the environment. If it were possible to establish early warning systems which accurately predicted for a farmer the onset and probable severity of an attack of a pest or disease, then he could rely entirely on remedial measures and apply pesticides only when they were really necessary. A great deal of research on this problem has been, and is being, done by Government, universities and industrial agricultural organisations, but the present state of knowledge is, as has already been noted, very far from providing a reliable comprehensive service. Some success in forecasting has been achieved where mass migrations of pests are concerned, thus, the Food and Agricultural Organisation of UNO have established very successful international early-warning systems for locust attacks. This, however, is a far cry from being able to predict the build-up of a pest or disease in a particular region, or even in a particular field. At the most, an indication may be given that a particular pest or disease is likely to be more prevalent in a particular season and this may simply predispose the farmer more strongly to adopt preventive measures. Disease and insect warning services are operated in many countries by advisory and extension services but there is scope for much more work in this area. One essential requirement is accurate long-range weather forecasting.

Even if a farmer knew that an attack of a certain pest or disease was imminent, he would still want to know what the severity of the attack was likely to be. If the attack is limited, it may make very little difference to crop yield and quality and the financial loss may not exceed the cost of crop protection treatments. On the other hand, the attack may spread with alarming rapidity and cause extensive damage before remedial measures can be applied and become effective. There is no method by which advisers can tell a farmer to go into a field and, if he sees so many insects on a leaf, to spray, but if he sees less, not to spray. In about 10% of cases it will be fairly obvious that extensive damage will occur unless something is done, and in about 10% of cases it will be reasonably certain that little damage will result but, in the 80% of cases in between, the farmer cannot be sure, so he will apply a pesticide just in case or, if he asks for advice from extension services, they will also tend to make the same recommendation.

It is possible for Government advisory services, manufacturers, food processors or growers' associations to collect and analyse data on the cultivation of a particular crop, in a particular region, over a number of years, since they have the facilities both to collect this information and to process it. By doing so, they can provide a useful back-up service to the farming community.

For example, the six agencies coordinated by the US Department of Agriculture have as their objective the provision of economic research on comparative farm costs and returns, and the collection of basic data on current practices, costs and methods of controlling pests, which involve use of chemicals in the major agricultural areas of the nation and on the effects of restrictions on the use of these chemicals in agricultural production. A very respectable amount of research has already been completed in these areas in the USA. Other developed countries also carry out similar work to a greater or lesser degree. In the UK the research establishments of the Ministry of Agriculture and the research units of the Agricultural Research Council such as the Weed Research Organisation and the Glasshouse Crops Research Institute have done, and are doing, very valuable work in these areas. Many university departments are contributing useful information towards these economic assessments and are pioneering new techniques of research. For example, the Environmental Resource Management Research Unit of Imperial College in the UK have applied statistical studies, systems analysis and computer modelling to an investment appraisal study of froghopper control in sugar cane. The study takes into account the biology and life-cycle of the pest and of its predators, and the ways in which populations build up and the extent of damage they cause, and relates these factors to economic losses. Different strategies of crop protection can thus be compared. A typical decision path which is used for investigation is shown in Figure 4.3. Similar studies are also being carried out by the Department of Industrial Science at the University of Stirling, also in the UK, on strategies of control of armyworm in lucerne production, by the University of Texas on boll weevil control in cotton, by the University of North Carolina on control of pests of tobacco and soya bean, and on other crop protection problems in universities in the USA, UK and other parts of the world. The food processors, because of their ability to collect and process a wide range of data with regard to specific crops, are also able to give pertinent advice on economical application of crop protection methods, as can many of the growers' associations. The larger manufacturers carry out a great deal of work in this area because it is relevant to the efficient operation of their businesses and also as part of their technical service to their consumers. Thus, in the UK, Plant Protection Ltd have carried out extensive studies on the economics of minimum tillage techniques, in which herbicide treatment to clear the ground prior to sowing is substitued for the conventional methods of cultivation such as deep ploughing, and also of the use of herbicides in improvement of pasture for grazing.

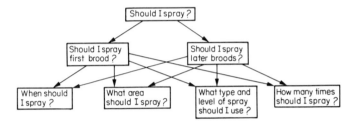

FIGURE 4.3 Decision for froghopper control. From G. A. Norton and G. R. Conway (Imperial College, England), *International Pest Control*, April 1974.

Nevertheless, the results of all such studies must be generalisations which apply to particular crops grown in particular regions. The individual farmer still has to make the decisions on crop protection —except when these are specified by food processors' contracts— which are appropriate to the total operation of his individual farm. He can be helped considerably by the provision of advice based on the conclusion of studies of the type described above which he can adapt to his own situation. The Government advisory and extension services together with the processors, the manufacturers and the associations have a vital role to play in the education of farmers in the economical and safe use of pesticides. Most modern pesticides present very little significant risk either to the consumer in terms of residues or to the environment and wildlife when used in the proper way. Most residue and environmental problems which do occur happen as a result of farmer error or ignorance. The US Department of Agriculture have done valuable work in the field of educational programmes and their success is reflected in the productions of American farms and forests and the excellent records of pesticide safety which have been established by American agriculture. In many other countries Government educational programmes have had a similar effect.

In summary, the questions with which farmers will be concerned in any comparative cost-benefit studies of various crop protection methods are:

1. What will be the financial effects on the total operation of the farm?
2. What will be the effects on labour requirements?
3. What will be the effects on the timing of farm operations?
4. What will be the effects on land requirements?
5. Will the yield and quality of treated crops be acceptable?
6. What will be the effects on personal convenience and leisure?
7. What will be the environmental effects, e.g. on pollinating insects?

5

The Food Processor's Viewpoint

In recent years the processed, convenience and frozen food industries have grown rapidly in developed countries (see Tables 5.1 and 5.2), mainly at the expense of the fresh food markets (see Table 5.3). The annual turnover in processed foods in the UK has increased twenty-fold since 1948 and it is estimated that, by the 1980s, half of all the foodstuffs

TABLE 5.1
Production of frozen vegetables in the UK.
(By courtesy of W. G. Kovachich, Unilever Research Laboratory, England.)

Year	Frozen vegetables (tonnes)
1956	23 936
1957	30 054
1958	31 568
1959	41 991
1960	57 600
1961	68 000
1962	79 800
1963	79 700
1964	85 800
1965	79 700
1966	97 600
1967	133 800
1968	118 900

TABLE 5.2
Potato processing in the UK (tonnes of raw materials). (By courtesy of W. G. Kovachich, Unilever Research Laboratory, England.)

Year	Crisps	Instant mash	Frozen	Canned	Soups	Fishcakes	Total
1955	50 000	15 000	—	500	30 000	20 000	115 500
1966	320 000	85 000	50 000	15 000	40 000	30 000	546 000
1970	525 000	125 000	100 000	50 000	50 000	40 000	900 000

TABLE 5.3
Types of peas sold in the UK. (By courtesy of W. G. Kovachich, Unilever
Research Laboratories, England.)

	1963	1965	1967	1968
Frozen	26%	29%	33%	36%
Canned garden	23%	20%	19%	18%
Canned processed	25%	25%	25%	24%
Fresh	26%	26%	23%	22%

sold will be in the form of prepared or convenience foods. This has largely been in response to changing social conditions. A large number of married women now have full-time jobs; they have not the time for the chores of cleaning and preparing vegetables and fruit and they have the money to spend to buy a greater variety of foodstuffs. The growth of supermarkets with their preference for pre-packed goods has contributed to this tendency. Many people live on housing estates out of town and make visits once weekly or less by car to stock the deep freezer which has become a reasonably cheap standard item of domestic equipment as a result of modern methods of mass production.

In the past twenty years this changeover to processing has radically altered the character of vegetable farming. There has been a massive movement from traditional growing regions to areas close to the processing factories. Cultural methods have changed extensively. Green beans which were traditionally grown on 2 to 5 acre plots and hand picked are now mainly cultivated in closely-spaced rows in large fields and are harvested mechanically. Green peas are now grown under very high density conditions and are shelled on the fields by fleets of mobile viners instead of being collected and taken to static vining stations. Whereas they used to be exclusively widely-spaced and hand-picked , many brussels sprouts are now close planted, topped and totally harvested mechanically. Potatoes are cultivated in special ways to produce large numbers of small tubers. Other vegetables have also undergone similar radical changes in farming methods. Fruit farming has been similarly affected by the increase in deep freezing and in pre-packaging. There is every indication that this tendency will increase in the future.

Intensive production has been achieved, and is maintained, by extensive use of fertilisers and crop protection chemicals including not only pesticides but also plant growth regulators, which are substances which can affect the size, shape, colour, time of ripening, etc., of fruit and vegetables. Moreover, as has already been pointed out, the consumer nowadays refuses to accept blemishes in processed and frozen vegetables and fruit. Before, such blemishes were considered quite normal in fresh produce. Achievement of this complete absence of pest or disease damage is very dependent on effective crop protection and pest control.

The processed food industry is highly competitive and its production costs have to be cut to a minimum by use of modern techniques of automation and mass production. The farm then becomes the first unit in a highly integrated production system and subjected to the same economic constraints as that type of operation. This has a number of consequences for the farmer, who must now regard his crop as a factory raw material. He will be encouraged by the processors to reduce his labour force to a minimum and to diminish his overheads. The processor also aims at increased efficiency by cutting the labour force which he himself employs for inspection of raw materials, and therefore demands a degree of freedom from extraneous materials, blemish and damage which implies that the farmer must grow his crops to very high standards. This tends to drive the small farmer out of this type of business and to encourage either large farms or growers' co-operatives who are able to invest heavily in mechanised equipment and thus reduce harvesting costs. The food processors are increasingly encouraging development of new high-yielding varieties of plants which, however, are not infrequently more susceptible to pests and diseases than earlier varieties. This fact, coupled with intensive growing conditions and abandonment of crop rotations, provides ideal conditions for growth and spread of such pests and diseases, necessitating stringent programmes of crop protection and increased reliance on pesticides.

Tropical crops such as tea, coffee and sugar-cane can be regarded as being grown mainly for the processed food industries but, because they are generally cultivated in regions where labour is cheap and plentiful, they are subject to different considerations.

In the processing of vegetables, damage by pests or diseases may not only make the final product unacceptable to the consumer but may also seriously interfere with the actual processing operations. Thus an incidence of more than 7% botrytis grey mould on beans makes the crop impossible to handle. Weeds present a particular problem: firstly, because they may interfere with mechanised harvesting programmes and operations and thus increase costs and, secondly, because they may introduce foreign matter into the final processed goods. For example, the whole of a year's pea crop has to pass through the processing factory within a period of a few weeks. The rate at which the crop can be harvested is governed by the total bulk of plant material not by the actual amount of peas, and so can be seriously slowed down by weeds. Dwarf beans are harvested by a picking reel which removes the pod from the plant and this equipment can be badly affected by tough branching weeds such as fat hen or by stringy weeds such as knotgrass and bindweed.

From the point of view of the processor a weedy crop is often a useless crop.

In addition to these operational problems there are legislative requirements relating to processed foods. In the UK the Pure Foods and Drugs Act demands that packaged foods must not contain any extraneous matter. In practice, small tolerances are permitted, but these may be as little as one piece of extraneous vegetable matter in each 10 kg of produce at the packing stage.

In order to be able to budget accurately, which is an essential for any economically viable factory operation, the food processors are increasingly contracting with farmers at the beginning of each season to buy the whole of their crop rather than trying to purchase on the free market at unpredictable prices. Naturally, such contracts will stipulate minimum quality requirements and, because of the considerations already described, these will be for substantial freedom from blemish, damage and extraneous vegetable matter. The consequences to the farmer of failure to meet such quality requirements can be severe financial loss since an unsatisfactory crop may have either to be ploughed in or sold very cheaply for animal feed. The farmer may therefore tend to use crop protection and pest control methods intensively to avoid the risk of this happening. This, however, poses additional problems for the food processors since they are legally responsible in many countries for ensuring that their products do not contain more than the legislatively permitted levels of pesticide residues. This means that the processor must either have a complete and detailed knowledge of all pesticides which have been applied to the crops delivered to him or else maintain a sizeable analytical laboratory to detect and measure residues of any types of pesticides. The costs of establishing and running such a laboratory are very high. It is obviously cheaper for the processor to write into the contract with the grower exactly what crop protection programme should be applied to the crop. The processor will tend to favour pesticides which are specific and which will be broken down in the plant so that no residues are left and, preferably, pesticides which can be applied as early in the crop's life as possible. They will not permit the use of any new pesticide unless the manufacturer can satisfy them on these points. A further point which has to be considered is that pesticides may introduce taint into processed foods which are much more susceptible to this problem than fresh produce.

Consequently, in fruit and vegetable growing, the decisions on crop protection methods are increasingly being taken out of the hands of the farmer. It is in the interests of the processors to obtain the co-operation of growers to detect and prevent any unauthorised spraying of crops. Even in the case of cereal crops, which, of course, are much larger tonnage crops than fruit and vegetables and occupy much greater acreages, the same situation is arising, either because wheat is grown under contract to the millers or barley under contract to the brewers, or, alternatively because high standards of quality and the use only of certain specific pesticides is demanded by these manufacturers for any crops which they are willing to purchase. In the UK the British Sugar Corporation contracts for all sugar beet grown and has a monopoly of supply of seed. Also in the UK in 1967 one third of all cereal seed grain was grown under contract to merchants.

This is in many ways a desirable tendency because it leads to a stricter control and much more complete information on the way in which pesticides are used. Furthermore, the food processors have both the expert staff and the computer facilities to collect and analyse extensive statistics, year by year, on all aspects of the crops with which they are particularly concerned, and most of them do this. They are well placed to co-operate with Government advisory and extension services and with chemical manufacturers to produce data on the incidence and spread of pests and diseases in various localities, on the effects of climate and other factors on these, and on the extent of crop losses. They can thus give advice to farmers on the most economical and safe ways of applying crop protection measures and regulate the ways in which these are used. Modern pesticides are subjected to such extensive toxicological and ecological studies before they are approved for sale by registration authorities that the risks of environmental ill effects are significant only if they are misused. Such misuse has occurred in the past, and may still occur, not because of unconcern on the part of farmers but because of lack of knowledge, that is, because of human error. Anything which helps to provide accurate advice on crop protection and pest control to the farming community is to be welcomed. The laying down of spraying schedules and designation of acceptable substances and quantities helps to avoid such errors.

However, food processors will aim at optimisation for a particular crop in a particular region since they are concerned with total economies and they will not be particularly concerned at the fortunes of individual farmers. The latter will, therefore, always have to exercise judgement in the operation of their own businesses. Nevertheless, contract growing will spread because food processing must develop further in order to meet the community's needs. The logistics of feeding the populations of large industrial cities are formidable and vast quantities of foodstuffs have to be moved in and distributed every day. It is difficult to see how this can continue to be done efficiently and economically without the aid of modern methods of food handling. One has only to look at the densely-populated cities of the Third World (the less privileged nations) to see the misery which lack of efficient food production and distribution can bring. Furthermore, deep freezing, pre-packing and similar processing methods will tend to eliminate the enormous post-harvest losses of crops which have always been associated with the fresh food market. The losses have been partly caused by decomposition and rotting during transport and distribution, as a result of fungal or bacterial attack, and partly by mechanical damage resulting from inefficient and careless handling in warehouses, on ships, in docks and at markets.

In summary, the questions with which food processors will be concerned in any comparative cost-benefit studies of various crop protection methods, are:

1. What will be the financial effects on the total operation?
2. What will be the effects on labour requirements?
3. What will be the effects on timing of operations?
4. Will the methods give the degree of control which is required?
5. Will they give rise to any residue problems?
6. Are methods of residue analysis available for residues which are simple and cheap to apply?
7. Will the methods used cause taint?

6

The Consumer's Viewpoint

The consumer wants food which is pure and wholesome, of high quality and attractive appearance, in as great a variety as possible, at reasonable prices and in good supply. The general increase in affluence of people in developed countries over the past thirty years has made it possible for them to get what they want since they can afford to pay for it. Apart from any of the other requirements mentioned above, the matter of purity and wholesomeness is one on which the consumer is very sensitive, as the fear of mysterious and secret poisons in the food he eats is a very deeply-rooted and atavistic one which prompts an exaggerated response if any suggestion of contamination is made. In previous centuries, widespread adulteration of foodstuffs was common commercial practice, but legislation and public inspection in developed countries have steadily brought this under control, so that any gross adulteration of food is now very rare. Legislation and general education in food handling and hygiene have reduced the likelihood of bacterial contamination of food, although there is still room for improvement in this area as local instances of food poisoning are still much too common. Nevertheless, public fears of gross adulteration or bacterial contamination of foodstuffs have been largely allayed. In recent years, however, the consumer has been made aware of what is to him a more insidious threat of the possibility of residues of pesticides in his foodstuff which he thinks might make him acutely ill or, more alarmingly, have long term effects on his health.

Government authorities responsible for registration of pesticides have, very properly and rightly, made provision of sufficient acceptable data on residues an essential requirement for permission to sell. Manufacturers have always been concerned about possible residual toxic effects of their products and have carried out toxicological studies of their effects on test animals and field studies of the amounts of pesticides actually present in harvested crops. With increasing public concern, these tests have been made more and more comprehensive, so that nowadays a new pesticide is subjected to an extensive and searching toxicological testing programme. This programme includes study of acute toxic effects from doses of various sizes; 90 day and two year continuous feeding to dogs and rats—after which the animals are humanely killed and every organ of their bodies closely examined histologically for any departures from normal; studies of reproduction over three generations of rats to ensure no effects on embryos and no transmitted effects to offspring; and studies of general toxicities to birds, fish and shellfish. In addition, extensive studies are made of possible environmental effects on beneficial and pollinating insects, on soil organisms and on wildlife in general.

At the same time it is a requirement for registration that analytical methods for a new pesticide shall be developed, capable of detecting and measuring 0·1 parts per million in meat, 0·01 parts per million in food crops and 0·005 parts per million in milk. Determinations then have to be made of residues present after harvest in all crops to which the pesticides may be applied and, in the case of animal feeding stuffs, of any residues in the meat or milk.

There has been a certain amount of confusion in the minds of consumers about residues. This is partly because the layman does not always appreciate that there are no poisonous or non-poisonous substances, only poisonous or non-poisonous amounts. The enormous increase in the sensitivity of analytical methods has increased this confusion. If the only analytical method available for a particular pesticide could not detect more than 1000 parts per million, then a sample containing 900 parts per million would be designated by the analyst as 'non-detectable', and the consumer would feel reassured. If then a method was developed for detecting 0·01 parts per million, a sample containing 0·03 parts per million would show up positive even though this is 30 000 times less than the amount in the previous 'non-detectable' sample. The consumer then feels uneasy on a general principle that 'if it is there at all, it must be doing some harm'.

This illustrates the crux of the pesticide residue problem. Everybody agrees that harmful residues should not be present, but what amounts should be specified as the maximum allowable for safety? We are up against the impossibility of proving a negative and of being able to say for certain that a particular amount of residue might not cause some harm to somebody, somewhere, sometime. Likewise, it is not realistic to specify that no residues should be present, because this can likewise never be proved, as the most that can be said is that the substance is 'not detectable' by the most sensitive analytical methods available. Registration authorities are forced therefore, into specifying amounts, that is, laying down finite limits for residues. The 64 thousand dollar question is, 'What is a safe amount?' To specify 'non-detectable by the most sensitive analytical methods' is unrealistic because modern analytical methods are so sensitive that the amounts involved are fantastically small—0·01 parts per million is equivalent to one teaspoonful in a rail tanker wagon. From an overall cost-benefit view, the banning of a pesticide, because it could not achieve such low limits, might have large economic repercussions on the production of one type of food, without any compensating benefit, since this action may have been taken against a non-existent danger. What then is a reasonable amount of residue to specify which will not expose the consumer to an unreasonable risk? What is an unreasonable risk?

A generally accepted basis is that the maximum permitted amount of residual pesticide in a foodstuff is determined from three factors: (1) the smallest dose, expressed in parts per million, which produces detectable harmful effects in experimental animals; (2) a safety factor which is usually 100, but may be less if a great deal is known about the physiological and pharmacological effects of the pesticide; (3) a food factor based on the proportion of the particular foodstuff in an average diet. Thus if the minimum harmful dose is 10 parts per million, the safety factor 100 and the food factor 0·2, the permitted residue level will be 10 divided by (100 × 0·2), that is 0·5 parts per million. The safety factor is intended to be large enough to compensate for any differences between man and test animals, for any variations in susceptibility due to age, health or personal idiosyncrasies, and for any differences in eating habits such as the person who eats an inordinate amount of a particular food. The maximum permitted residue levels for pesticides differ from country to country and this causes difficulties in international trade, because a foodstuff which is acceptable in its country of origin cannot be exported to a country where it is not acceptable. It would be simpler if nations could agree on world standards and, in fact, the Food and Agricultural Organisation and the World Health Organisation of the United Nations have published recommended residue limits for most commonly-used pesticides in most commonly grown crops, which they consider would not expose the consumer to appreciable risk on the basis of known facts. However, cost-benefit considerations may show that what is a reasonable residue limit for one country may not be reasonable for another which has quite different circumstances. Unnecessarily low residue limits in developed countries create difficulties for those developing countries whose economies are dependent on export of agricultural and horticultural produce and who cannot afford very expensive methods of crop protection and pest control.

Possibly some guide to acceptable margins of safety may be obtained by comparing the toxicities of pesticides with those of naturally occurring toxicants which are found in food and foodstuffs both of men and animals, since this is the background against which their potential hazard must be judged. By and large, toxicities of naturally occurring substances in foodstuffs extend over a wider range than those of pesticides and nature has created many substances which are far more toxic. Small amounts of a great variety of highly toxic natural products are widely dispersed in the environment. Thus the average fatal doses for a 70 kg man of some poisons which may occur in common foods are as follows: in mouldy cereals (aflatoxin) 0·38 g, in mouldy rice (islanditoxin) 0·45 g, in bad fish (tetraditoxin) 0·0007 g, in contaminated shellfish (saxitoxin) 0·0003 g, in toadstools (amatoxin) 0·07 g. By comparison, the corresponding values for a number of commonly-used pesticides are: DDT 21 g, dieldrin 3·5 g, malathion 42 g, parathion 0·3 g, 2,4-D 35 g, atrazine 210 g. Parathion is probably the most toxic pesticide in general use.

The problem of determining safety is a problem of interpreting data obtained on experimental animals under laboratory conditions in terms of an actual hazard to people in practical use. Nobody is sure how to do this. Is there, in fact, any evidence of harm caused by pesticide residues in practice? In developed countries regular 'market-basket' surveys are made in which Government analysts purchase various foodstuffs in the normal way in shops and supermarkets and analyse them for pesticide residues. Such surveys invariably show that most samples contain well below the acceptable levels of residues—for instance, the Government analyst in the UK has reported that residues were, in general, of the order of a few parts per thousand million in the various foodstuffs. In the few cases where abnormal levels of pesticide residue could be detected, these were traced either to human error or human greed. For instance, a recent case in the southern United States in which millions of broiler chickens had to be destroyed because they contained more than the acceptable amount of an organochlorine insecticide, resulted from the fact that seed treated with this substance had been improperly obtained and sold as animal feed by unscrupulous operators.

The very low levels of pesticide residues found in practice are not surprising when it is realised that the acceptable residue levels are not the amounts which normally occur, but the maximum amounts which could result from use of the pesticides in a proper way. They serve, therefore, as a means of detecting improper use of pesticides. Instructions for use and applications are drawn up by the manufacturers so that, if these are followed, there is no danger of undesirably high residue levels.

Most attention has been given to residues of the organochlorine insecticides such as DDT, γ-BHC and dieldrin, since these have been used on a large scale on a variety of crops in many different areas. In Table 6.1 there is shown the average total intake of these compounds for a 70 kg man in the UK, based on figures obtained for residues in all types of foodstuffs from extensive market-basket surveys, compared with the maximum daily amounts which are regarded by the World Health Organisation as hazardous to a man if eaten every day throughout his whole life. In the words of the UK Government analyst, these figures are 'reassuringly low'. Only in the case of dieldrin does the observed intake come anywhere near the WHO figure and use of this compound has now been greatly curtailed by voluntary agreement with the manufacturer. The actual average amounts of DDT, γ-BHC and dieldrin observed in

TABLE 6.1

Average daily intake of organochlorine insecticides in the UK in 1967. (From Third Report of Research Committee on Toxic Chemicals, Agricultural Research Council, England.)

Insecticide	Intake from total diet (micrograms/day*)	WHO acceptable daily intake (micrograms/day*)
BHC	6·6	875
Dieldrin	6·6	7
DDT	44·4	700

* 1 microgram approximately equals one 500 millionth part of a lb.

TABLE 6.2

Organochlorine insecticide residues in food in the UK, 1966–1967. (From Third Report of Research Committee on Toxic Chemicals, Agricultural Research Council, England.)

Food group	BHC (parts per milion)	Dieldrin (parts per million)	DDT (parts per million)
Cereals	0·014	0·003	0·018
Meat	0·016	0·009	0·050
Fats	0·059	0·024	0·208
Fruit	0·005	0·002	0·026
Root vegetables	0·004	0·002	0·006
Green vegetables	0·006	0·003	0·012
Milk	0·003	0·002	0·004
Whole diet	0·010	0·004	0·028

TABLE 6.3

Organochlorine insecticide residues in food in the USA in 1965. (From J. G. Cummings, *Residue Reviews*, 1966, **16**, 30.)

	BHC (parts per million)	Dieldrin (parts per million)	DDT (parts per million)
Amount in total diet	0·010	0·076	0·091
No effect level in long-term feeding studies on animals	25	1	500

various actual foodstuffs in the UK in 1966–7 are shown in Table 6.2. Figures for the United States are given in Table 6.3. These figures should be considered in relation to the fact that the smallest single dose of DDT which will produce clinical symptoms of any kind in an adult man is about half a gram and up to 20 g can be taken in a single dose without death.

The general conclusion is, in the words of Dr J. M. Barnes, Director of the UK Medical Research Councils' Toxicological Research Unit, 'that there is not a tittle of scientific evidence that pesticide residues in human food present a real threat to health'. The organochlorine compounds are the only pesticides which are detected regularly and the general levels observed are well below the accepted safety levels. Other pesticides are detected only very rarely and then in very tiny amounts. Even the very widely used organophosphorus insecticides are detected in only about 1 in 70 of samples of food tested, and even then at exceedingly low concentrations.

There remains the contention that is sometimes put forward that, although the trace amounts of pesticides actually present in foodstuffs will not cause any acute damage to health, they might have subtle long term chronic effects. In this connection, cancer is a very emotive word. It is very difficult to refute this type of argument because it involves conclusively proving a negative. The argument that compounds which have been shown to have no detectable toxic effects in very small amounts might be harmful in trace amounts is an odd one. Experience with drugs has shown that, in very small amounts, many dangerous substances can act as beneficial stimulants.

On the question of cancer it is certainly true that some chemical substances have been unequivocally shown to promote the formation of malignant tumours, but none of these are used as pesticides. There is, however, a very much larger number of very common chemical substances which can be induced to promote tumours in susceptible experimental animals by implanting massive amounts of them under the skin. These include such common substances as gold, which has been worn in contact with human skin for many years without any recorded ill-effects. It is difficult to assess what meaning experiments of this type have in relation to an actual risk in practice. Some pesticides have been suspected of producing tumours in animal experiments carried out under these highly artificial conditions but this is not necessarily a good reason for restricting their use. Most attention has been focussed on DDT because this is a highly persistent chemical which tends to accumulate in body fat so that nearly all people have about 3 parts per million of this compound in these fats. No evidence has yet come to light of chronic (i.e. long-term) poisoning in man by DDT. Workers in DDT factories in the USA have absorbed about 35 mg per day for many years and have remained healthy and free from clinical symptoms, and American volunteers who ate this amount daily for two years suffered no detectable ill-effects. The problems of possible subtle long-term effects of continuous ingestion of extremely minute amounts of pesticides is one which requires more investigation and study.

It is obviously desirable for growers to use pesticides which, on the basis of toxicological investigations on animals, are of as low toxicity as possible so that the safety margin is wide and, preferably, they should not be persistent but be broken down rapidly in practical conditions to harmless substances. These requirements are more than ever being taken into account in development of new crop protection chemicals.

In summary, the questions which will concern the consumer in any comparative cost-benefit study of various crop protection treatments are as follows:

1. What will be the effects on price of foodstuffs?
2. What will be the effects on quality and variety and availability of foodstuffs?
3. Will there be any detectable pesticide residues and, if so, what is the risk that they might be harmful either short-term or long-term?

7

The Public's Viewpoint

The general public's concern over pesticides (apart from their attitude as consumers of food) centres on whether use of these chemicals in any way affects the quality of their lives. Their attention was first directed to the possibility that widespread use of crop protection chemicals might have adverse effects on plant and animal species which were not intended to be the direct targets of the treatment, and that the quality of the environment might consequently be diminished, by a book *Silent Spring* which was written by a US journalist, Rachel Carson, and published in 1963, and by the determined campaign which Miss Carson carried out subsequently. This book was, unfortunately, not a scientific discussion of the problems but an emotive appeal which was subsequently taken up and extended by the mass media, and resulted in the emergence in the United States of a powerful, well-financed anti-pesticides lobby.

This turn of events is to be regretted. Right from the start of the modern pesticides industry, ecologists, biologists and chemists, both outside and inside the chemical industry, have recognised the possibility that pesticides might affect non-target species. In his book *The Scientific Principles of Plant Protection* published in 1928, Dr Hubert Martin, one of the most eminent pesticide chemists, makes this clear. Rachel Carson was a biologist and the basic dangers to which she drew public attention are real dangers which deserve careful consideration. The indiscriminate public reaction against pesticides which has been stimulated by distorted presentation of these dangers has had the unfortunate effect of forcing those intimately concerned with production and use of pesticides to take up defensive positions so that there is, in the United States, an apparent confrontation between the environmentalists and conservationists on the one hand and the chemical manufacturers on the other. This is unnecessary as, fundamentally, there is no reason for conflict between them. Everybody concerned with crop protection, whether they are conservationists or manufacturers, accepts that use of chemical pesticides poses environmental problems which are extremely difficult and complex to solve, and that it is essential to work together to amass the vast amount of scientific information which will enable them to be solved and permit rational decisions to be made with regard to registration requirements and regulatory control measures. Only in this way will it be possible to approach an optimisation of cost-benefit of crop protection for the community as a whole. Overstatement always detracts from the force of an argument and some of the more unsubstantiated accusations made against pesticides have devalued those matters which really are of significance and importance. It is noteworthy that, in the UK, where there has been much less public reaction to pesticides, the organisations concerned with environment and conservation, the Government Organisations concerned with regulatory controls and the pesticide manufacturers all work together voluntarily in a spirit of co-operation to try to reach sensible decisions based on a scientific assessment of risks.

A joint British Agrochemicals Association/Wildlife Education and Communication Committee representing manufacturers, distributors and users of pesticides, the Ministry of Agriculture, the Nature Conservancy and various other Conservation Organisations together produced *Pesticides: A Code of Conduct* in 1968. This represented an agreed view on the responsibilities of all the various types of people concerned with pesticides to ensure that they are used beneficially without harm to humans, to wildlife or the environment in general, and on the ways in which this should be done. The extent of co-operation in the UK is reflected in the fact that the Government can rely on a voluntary rather than legislative control of registration, sale and use of pesticides. Possibly this is because intensive use of pesticides is not normally required under UK conditions. The book *Pesticides and Pollution* by Dr Kenneth Mellanby, the Director of the Monks Wood Experimental Station of the British Nature Conservancy, is recommended as the calm and balanced views of a man who is deeply concerned both with wildlife and its preservation and with human welfare.

It needs to be recognised and accepted by all those who are striving to solve the problems of pesticides that mistakes were made in the early years of modern crop protection, but that such mistakes and mishaps are an inevitable aspect of the development and application of any new technology. Such errors are made, not because of greed or malice or lack of concern, but nearly always because of ignorance. These early mistakes should be openly accepted for what they were, neither played down nor minimised, but also not exaggerated. In the case of pesticides, farmers were inadequately warned about the inherent hazards of distributing highly biologically active chemicals over wide areas, and instances of misuse of pesticides occurred. There is, nowadays, a much greater awareness by farmers of the need to use pesticides wisely, although there is still scope for further education. Organisations concerned with discovery and development of new pesticides have steadily expanded their environmental and ecological studies groups, so that nowadays a much greater amount of information can be passed to registration authorities. It is of vital importance to the future well-being of mankind that society makes use of crop protection chemicals safely and wisely if there is to be any hope of satisfying man's health, food and fibre needs. At the same time, every effort must be made to estimate undesirable side-effects. There is a need for all concerned to work together to win the war against man's natural enemies without devastating the battlefield, and not to waste effort fighting amongst themselves.

Why does wildlife matter? A large proportion of the general public, which has often responded in a highly emotional way to propaganda against pesticides, would give a very superficial answer to this question, probably in terms of fears that there might be fewer wild flowers to pick or wild animals and birds to see or hunt when they go into the countryside. For the environmentalist and conservationist, and for those members of the public who have some insight into these matters, the answer is a much more profound one. They would say that life is such a mysterious and wonderful phenomenon that we should respect all its many forms, not just from sentiment or because of the aesthetic appeal of flowers, birds, wild animals and wild places, but because man has subtle inter-relationships with all other forms of life and should be very cautious about how he interferes with or affects these. These living organisms have evolved over millions of years, during which time various species have become extinct through changes in their environment and many species have evolved; but it may be unwise for man to introduce environmental stresses which accelerate this process to such an extent that the natural system has insufficient time to adjust and severe imbalances arise. We do not know enough to be able to predict with certainty what the results could be. The surface of this planet forms a complex system of which man is but one part and with which he must learn to live in harmony.

From the purely economic point of view, the financial viability and standard of living of the population of many regions nowadays depends on income from tourism, so that preservation of natural amenities is a matter of vital concern to them. Also, from a social as well as an economic viewpoint, it is important to preserve an adequate genetic reservoir of wild animals and plants on which man can draw in the future for new varieties of domestic animals and cultivated plants to rectify any shortcomings which may develop in those at present in common use—for example, as a result of increased susceptibility to mutant strains of pests and diseases, or which can provide new sources of food for a rapidly increasing world population.

Environmentalists and conservationists recognise that man has constantly had to interfere with nature in order to maintain his species and that he could not for one moment relax his perpetual struggle against natural enemies of himself, his animals and his sources of food. They do not call for wholesale banning of pesticides, the value of which they fully recognise, but they seek legislative controls which are adequate, if fully observed, to prevent improper and indiscriminate use. They wish to have a proper understanding of any risks which pesticide use might involve, and they ask for consideration always of these risks in relation to what is achieved.

This is a view to which those who are concerned with the discovery, development and manufacture of pesticides fully subscribe so that, basically, there is no real conflict between them and those who are primarily concerned with the environment. Both advocate the cost-benefit approach which is the basic theme of this book, that is, the weighing up of all the advantages and all the disadvantages against each other. Differences of opinion may arise in quantifying the various aspects of the total cost-benefit balance sheet, but there is no reason why these cannot be resolved. In his Preface to *Silent Spring*, Julian Huxley has the following to say about ecology, which is the science that deals with the relationship of living organisms to their environment:

> Ecology cannot be merely quantitative or arithmetical; it has to deal with total situations and must think in terms of quality as well as quantity. One conflict is between the present and the future, between immediate and partial interests and the continuing interests of the entire human species. Accordingly, ecology must aim not only at optimum use but also at optimum conservation of resources. Furthermore, these resources include enjoyment resources like scenery and solitude, beauty and interest, as well as material resources like food or minerals; and against the interest of food production we have to balance other interests, like human health, watershed protection and recreation.

This is a reasonable statement of the systems approach to cost-benefit analysis of crop protection.

What then is the evidence that pesticides have caused damage to wildlife? Nearly all such evidence relates to one small group of pesticides, namely, the organochlorine insecticides which include the widely used compounds DDT, aldrin, dieldrin and lindane. There is some evidence also against a few of the many organophosphorus insecticides. This is not to say that other pesticides could not, if misused, have damaging effects on wildlife, and it is essential that a constant watch be kept for such damage. However, there is no evidence that any has occurred and the extent of ecological and environmental information which now has to be collected and supplied to registration authorities makes it unlikely that any new pesticide would produce such effects, if properly used in accordance with the manufacturer's instructions. Moreover, herbicides comprise about 70% of the total sales of pesticides and they are, by and large, the least poisonous and least environmentally hazardous of all crop protection chemicals. In his book, Dr Kenneth Mellanby says

I am of the opinion that weedkillers are not a major danger to wildlife. For most purposes we are using less toxic and less persistent substances. On the comparatively small areas where no plant growth is tolerated we use persistent substances, but these are apparently almost non-poisonous to animal life and do not 'leak' from the places to which they are applied. Wrongly or carelessly used, herbicides can obviously do a great deal of harm, but as a rule they make no more drastic changes to the environment than do normal mechanical farming practices—and some recently discovered chemicals may actually have less effect, for instance, on soil fauna.

Although fungicides are often very poisonous, most of them cause little obvious damage to wildlife. Already less toxic and less persistent chemicals are being used, and the tendency is to reduce rather than increase doses, even though a larger area of the country is treated each year. Some caution is, however necessary.

To deal first of all with the organophosphorus insecticides. In the early days of modern pesticides between 1950 and 1960 the organophosphorus compounds used were very poisonous to man and animals and special safety precautions had to be adopted to apply them. Many bird deaths, particularly of pheasants and partridges were a result of contact with freshly-sprayed plants. Since the organophosphorus insecticides are not very persistent—that is, they do not last very long in the soil or accumulate in insects or animals—it is doubtful if these deaths, although they caused intense public reaction, had any significant long-term effects on total populations especially as the effects were mainly on game birds which were destined to be shot in any event. In recent years, the most poisonous organophosphorus compounds have been gradually replaced by new compounds which are much less toxic to warm-blooded animals so the dangers from this group of compounds are being steadily reduced.

The earlier evidence against the organochlorine insecticides is quite substantial. In Britain in the spring of 1960 and of 1961 a large number of bird deaths occurred as a result of eating seed which had been treated with dieldrin, aldrin or heptachlor since the birds often take grain at that time of the year when other food is scarce. Large numbers of game birds were killed as also were predatory animals such as foxes and predatory birds such as the peregrine falcon which fed on the bodies of contaminated birds. The trouble was caused by incorrect and, sometimes, thoughtless use of seed treatments. Following these episodes, the manufacturers entered into a voluntary agreement greatly to restrict the use of such seed treatments and, as a result, there have been no similar episodes since.

In the United States there were, in the 1950s, similar problems with seed-treatments but there were also some unfortunate happenings as a result of treating very large areas of land with organochlorine insecticides, and it is these which are presented in dramatic terms in *Silent Spring*. During this time organochlorine insecticides were applied from the air over very large areas in various parts of the United States in operations designed to control, for example, Japanese beetle, gypsy moth and the fire ant. These operations resulted in the deaths of birds, small mammals, domestic pets and fish. It is doubtful whether any attempt was made at a cost-benefit analysis of these operations prior to their implementation, otherwise serious consideration would have been given to the question whether the results which were likely to be achieved justified the risks that were likely to be run. In the event, damage was done to wildlife and the control of the insect pests achieved was not very impressive.

It is generally accepted nowadays that any operation in which large areas of land are to be sprayed from the air with pesticides should be supervised by ecologists and biologists whose job is to ensure that the treatments are likely to achieve the desired results and that the hazards to wildlife are minimised. The extensive environmental and ecological work which now has to be carried out on any new pesticide makes it very unlikely that incidents involving widespread destruction of wildlife could occur again unless a user grossly disregarded all of the manufacturer's instructions. The safeguard against such misuse—which is mainly caused by ignorance—is the increasing education of farmers in safe and effective ways to apply crop protection treatments. The only safeguard against misuse from lack of concern is the legislative sanctions which are applied to anybody who commits an anti-social act, and laws against misuse of pesticides should be stringent. This should apply not only to people who cause excessive damage to wildlife by flagrant misuse but also to people who discard unwashed pesticide containers or leave pesticides in unlabelled and improper containers within reach of children.

The danger of large scale episodes of wildlife destruction has, therefore, been largely removed. It is doubtful whether such incidents had any permanent long-term effects on wildlife populations and, in most cases, recovery has been much more rapid than ecologists anticipated. It remains to examine the question whether pesticides could have any more subtle effects on wildlife which could have lasting effects on populations of mammals, birds and beneficial insects. In the book *Pests of Field Crops* by the two eminent biologists, Professor F. G. W. Jones and D. M. G. Jones, the matter is admirably summarised as follows:

> The possible accumulation of persistent insecticides of the
> organochlorine group in the fatty tissues of animals and their

progressive concentration up the food chain has given rise to much concern amongst naturalists and those interested in the preservation of rare or relatively rare species, but there is little acceptable evidence that organochlorine or other pesticides accumulate in this way. Failure of some predatory birds to breed successfully has been claimed to be caused by organochlorine compounds. Since 1950 more egg breakages have been recorded in the nests of peregrines, sparrowhawks and golden eagles and the relative weight (largely thickness) of eggshells has decreased since 1948. However, these phenomena appear to have preceded the use of organo-chlorine pesticides. The greater use of organic chemical compounds, particularly organochlorines, for domestic veterinary, agricultural and horticultural use began, if anything, after the thinning of the egg shells first appeared, so the causal effect is debatable although it is often accepted as true. Many sensational claims that insecticides have caused the deaths of birds and other animals are made in the press, on radio and television, often without objective evidence; condemnation precedes evidence because it supports the popular viewpoint although what proportion of deaths can rightly be ascribed to pesticides is uncertain.

It should be remembered that all animals and birds die eventually and that it is patently untrue to assume that every observed corpse is the result of pesticide poisoning.

In the UK a Committee on Toxic Chemicals was set up by the Agricultural Research Council and reports regularly. In its third report published in 1970 the Committee gave great attention to the organo-chlorine insecticides. They said

Whatever may be the physiological actions of novel materials present in the environment, if they do not reduce, or contribute to the reduction of, population sizes or affect population, they are not threatening to the conservation of nature. The final test of this can be provided only by long-term nation-wide ecological studies of populations. In nature, when a significant change in numbers is detected in a particular species, the cause of the change may not be clear and it would be quite wrong to attribute all changes to pesticides. In general, the most potent causes of population changes are changes in their environment, many of them man-made, such as removal of hedges, removal of farm ponds, spread of towns and motorways, trimming of road verges and reduction of small woods and copses and also agricultural practices such as ploughing and cultivation. Weather is also important, for example, the severe and prolonged winter of 1962–1963 greatly reduced the numbers of many species of birds for a few years. Localised incidents of pesticide misuse are not likely, in general, to affect a species as a whole. Such incidents are fairly readily recognised and ought to be controlled. On the other hand, the widespread presence of a material in or through the environment might not be so readily controlled; if it were to become dangerous, it might affect a whole species. Certain persistent organochlorine insecticides are probably present practically everywhere in the world, but their concentrations are at present too low to cause damage. Nevertheless, we ought to understand the position, and research should be encouraged to discover what these concentrations are, how they reach their present situations, what effects they have on living things and what can be foreseen about their future.

The situation with regard to pesticides and effects on wildlife can, therefore, be summarised as follows. Because of ill-considered application, a number of incidents of destruction of wildlife occurred in the early days of use of organochlorine compounds—although not in the UK—but there is no evidence that these incidents have had permanent long-term effects on population numbers. There is a suspicion that, because the organochlorine insecticides are very persistent and tend to accumulate in biological food chains, the reproductive capacity of some predatory birds may have been affected, but this is by no means proved. There is no evidence that any other pesticides have produced long-term effects on wildlife populations.

A rational attitude based on these conclusions would be to continue to require adequate environmental and ecological studies as a necessary part of registration requirements, to keep a constant watch for any adverse effects on wildlife, to supervise carefully any application of pesticides over large areas, particularly from the air, and to avoid the use of highly persistent pesticides when some acceptable alternative is available. This last point is prompted by the impossibility of ever conclusively proving a negative. Although there is no evidence that persistent pesticides are doing harm through their low-level presence in the environment we cannot be sure that no harm, however small, is occurring nor that they will never be concentrated to harmful levels by some living organisms, so that, when use of persistent compounds can reasonably be avoided, it should be. The decision must, in each case, be made on a cost-benefit assessment. If use of a particular persistent pesticide is essential for protection of a particular crop and there is no effective alternative it may be that a certain amount of risk of environmental damage may have to be accepted by society as the price it pays for the benefits it receives. It has been said that 'the perfect is the enemy of the merely good'. To refuse to use pesticides which are of overall benefit to mankind because they have some shortcomings would not be sensible if it led to food shortages while the search for the perfect pesticide was in progress.

A great difficulty is the complexity and uncertainty of environmental and ecological work. A vast amount of data have been collected, and are being collected, on the concentrations of pesticide residues present in various parts of the environment using highly sensitive analytical techniques which are often capable of detecting as little as one part of pesticide per million million, equivalent to 1 gram distributed through 1 million tonnes. The concentrations of pesticides found are generally very small, but there is no method at present available by which this vast amount of data can be interpreted in terms of a quantitative risk to any species of wildlife. A practical answer may be to maintain constant ecological watch for any detectable effects on wildlife populations and, in the absence of these, to assume that any risk is hypothetical.

Two particular parts of the environment merit special consideration, namely, the soil and water, including watercourses, rivers and the sea. Much work has been done on the environmental effects of pesticides in the soil. Pre-emergent herbicides occasionally persist in sufficient quantity to damage succeeding crops. Insecticides can cause considerable damage to arthropod populations in soil but there is little evidence to suggest that their effects are long-term and that fertility is thereby impaired. Fungicides and nematicides generally improve soil productivity by controlling plant diseases and have only occasionally been reported to have short-term detrimental effects on crop growth and yields. In general, the conclusion is that most pesticides cause only the type of impact on the soil environment which is intended and inherent in their use and no more than is caused by cultivations or crop rotations.

Water is of particular importance because it is a medium by which persistent pesticides could become widely distributed. Once again most attention has been focussed on the organochlorine insecticides because these have been used in the greatest amounts all over the world. Both air and rainwater contain minute but measurable amounts of these substances. For example, the average content of rainwater in the UK is about one part in ten thousand million. The average content of UK rivers is about eight parts in one hundred thousand million. It has not been possible to detect organochlorine compounds in the sea, even with the most sensitive analytical methods, although they must be there as they can be detected in the bodies of fish and marine animals. This is because many of these animals can extract and concentrate in their bodies the minute amounts of these chemicals, and this could conceivably cause them harm. There is need for more research on how pesticides get into waters, what effects they produce on life in the waters and what eventually happens to them, and on the actual toxic effects of various pesticides on freshwater and marine life. Most of the wilder accusations that have been made—such as the suggestion that DDT in the sea could reduce photosynthesis in marine phytoplankton and so reduce the oxygen content of the air—are not supported by experimental evidence and, in general, it would seem that, so far as pesticides in water are concerned, we are probably within the limits of safety except in places where massive contamination has occurred by lack of control, for example, by uncontrolled discharge of effluents from factories. Nevertheless, until much more precise knowledge can be obtained, it is sensible to be particularly cautious about the use of pesticides near watercourses, to avoid the use of persistent pesticides wherever possible, to maintain constant surveillance on rivers and on freshwater and marine life to detect any adverse changes and to ensure that the toxic and ecological effects of new pesticides on fish and marine animals are carefully studied before they are sold. All these things are, in fact, now being done in most developed countries.

In summary, the questions which will concern ecologists and conservationists in any comparative cost-benefit study of various crop-protection treatments are as follows:

1. What will be the effects on non-target species and wildlife in general?
2. Are there likely to be any subtle long-term effects on the populations of any species?
3. Are the pesticides persistent?
4. What are the transport mechanisms involved and where are the environmental risks?
5. Do the advantages of the treatments greatly outweigh any possible effects on the environment?
6. Are there any safer alternative methods of achieving the same degree of protection or control?

8

The Government's Responsibilities

A primary responsibility of a Government is to manage the Country's economy in such a way that all the population are at least provided with adequate food, clothing and shelter. The social consequences of severe food shortages in urban communities would be very damaging. Over and above this, they should encourage the development of manufactures and the application of new technologies to increase the real wealth of the nation and thus permit the material standards of living of the inhabitants to rise in an equitable manner, subject to the constraints of rewarding individuals in accordance with their service or labour. They should aim to do this within a system of laws and regulations which allows maximum freedom to individuals to live and enjoy their lives in their own ways, provided that they do not cause annoyance or harm to their neighbours. The Government should also try to safeguard the quality of people's lives as well as their material needs, first and foremost their health and, secondly, their more subtle aesthetic, social, recreational and religious desires and aspirations. All this should be done without mortgaging the well-being of future generations, and short-term gains should always be balanced against long-term losses. The Government is the manager and trustee of the nation's natural resources of all kinds. No Government could, in practice, ever achieve all the ideals outlined above and the best that they can hope to do is to provide rough justice by taking action to mitigate the hardship or distress which social or industrial developments may cause from time to time to specific sections of the population. But this should not detract from the value of these ideals as an objective. In the real world, a certain amount of abuse and injustice have to be tolerated as the price of freedom since the alternative is to regulate every detail of an individual's life to an intolerable and unacceptable degree.

A second responsibility of Government is education—that is, making people aware of the ways in which technological developments can be used effectively, economically and safely, and of the ways in which natural resources can be conserved and not squandered. The advisory and extension services in most developed countries are increasingly doing this for farmers.

A third responsibility of Government is to enact laws and regulations—and to provide means of enforcing them—which will enable individual actions which are considered contrary to the interests of the community to be restrained, and to provide means of redress and compensation when the careless, negligent or inconsiderate acts of one individual cause harm, distress or loss to another. There will, of course, be a wide range of different opinions amongst members of the public as to what constitutes an action which is contrary to the interests of the community, and the powers of decision on such matters are delegated to the politicians who constitute the Government and who will be influenced by their own opinions and interests, by the advice of their civil servants and, to a greater or lesser extent by representations made to them by specific pressure groups. In a democracy, a political party whose decisions run counter to the opinions of a majority of the population can be replaced at election time.

It is important to realise that laws and regulations can never prevent abuse but merely act as a deterrent, and provide a means of redress to it if it occurs. Many useful things contain within themselves the potential for harm if misused, and this possibility has to be accepted as the cost for their usefulness. Knives, for example, may be used to wound or kill and they can cause damage if carelessly handled, but their usefulness is such that nobody would suggest discontinuing their manufacture and use for this reason. Likewise, use of electric heaters should not be forbidden because an act of negligence with one could conceivably start a fire which might burn down a city, but every attempt should be made to make them as safe and as foolproof as possible.

From these simple examples it will be appreciated that a very large number of things which people use and actions which people take present a cost-benefit problem. A needless risk for which there are no compensating benefits should never be accepted but if, as in most cases, there are benefits, then these have to be assessed and balanced against the concomitant risks. Sometimes the risks are such that use of the particular thing has to be restricted to experts as, for example, administration of dangerous drugs is restricted to doctors. In the case of technical products, the political decision-takers should try to achieve an overall view of cost-benefit to the community as a whole. This is a heavy responsibility and they should, if they are to fulfil their duty to the electors, lean heavily upon the advice of technical experts in Government organisations and elsewhere and refuse to be unduly swayed by sectarian pressures.

It is interesting to compare the attitude of the public and of the politicians to two technical products, (1) cars, (2) pesticides, and to consider how each might be dealt with from a cost-benefit viewpoint.

The benefits of the car (including trucks, lorries, etc.) are that it makes possible the complex and rapid transport of raw materials and distribution of goods of all types which are essential for the efficient operation and economic viability of a modern industrial community. It provides rapid personal transport for business and social purposes and permits workers to live in widely dispersed catchment areas around their places of employment. It also provides recreational opportunity and allows large numbers of people to get away from urban areas and visit the countryside and the sea. In its manufacture, the car provides employment for many people and, in most developed countries, contributes a large amount to the gross national product and to the nation's export trade. It stimulates the ancillary business of garages, repair shops, motels, cafes, etc.

The social costs of the car are that it utilises large amounts of irreplaceable raw materials in its manufacture and vast quantities of limited world supplies of oil both in its various stages of production and in its operation and use. It causes atmospheric pollution, particularly in urban areas, and noise pollution in country towns and villages on main highways and in cities. It kills and injures large numbers of people and animals every year. By the demands it has created for highways, parking lots, etc., by the spread of urban housing development it has permitted, and by the access it has given to large numbers of people to the countryside it has indirectly reduced the amount of agricultural land available and has caused far-reaching changes in the environment and in the flora and fauna of the nation.

The Government's attitude has been to accept the cost of producing and maintaining adequate road systems with all necessary safety devices such as road markings, traffic lights, etc., to require that vehicles shall be roadworthy, to insist on prescribed tests before a person is licensed to drive, to lay down systems of laws and regulations covering the use of motor vehicles, and to provide a body of police and highway patrols to detect careless or negligent handling and infringements of any of the legal requirements. More recently there have been moves towards Government control of features of car design to limit pollution emitted from exhaust systems. All this is financed partly as a charge on general taxation and partly by specific taxation of cars and their users.

The attitude of the public is conditioned by their awareness of the personal benefits of their own cars which they would not willingly forgo or have severely restricted or curtailed. This results in considerable opposition to any moves to restrict or control access of cars to city centres or to areas of the countryside. There is also a general unwillingness on the part of the public to pay more for their cars or their motoring in order to reduce pollution and environmental effects. The very considerable social costs of the car are tacitly accepted because it is generally considered that the benefits justify their acceptance, even when human lives are a part of the costs.

The benefits of pesticides have been outlined in this book. They make possible production of adequate amounts of wholesome and nutritious food at reasonable prices, they help protect public health and control the discomfort of noxious or annoying insects, weeds, etc., they assist in the management of recreational land and water, they economise in the use of scarce agricultural land, they effect savings in use of scarce labour, their use produces savings in energy utilisation especially of fossil fuels such as diesel fuel for tractors, they assist in satisfying the public's fibre needs, and they protect raw materials and manufactured goods from biodeterioration. Like the car, their manufacture provides employment and contributes to the gross national product and to export trade, and their use has enhanced the economic well-being of the agricultural community and increased agriculture's contribution to the gross national product.

The social costs of pesticides are a very small number of deaths and injuries caused by accidental or deliberate ingestion, the occasional possibilities of harmful amounts of toxic residues in foodstuffs as a result of misuse and slight risks of harm to the environment and to wildlife if they are used in an improper or negligent way.

It is interesting to note that in 1965, total accidental deaths from pesticides in the USA were 356, about the same number as died in road accidents on a one day national holiday in that year, and that 70% of the deaths from pesticides were caused, not by modern products, most of which are relatively non-poisonous, but by pesticides introduced before World War II.

It is suggested that the policy of Government should be to require new pesticides to fulfil certain prescribed requirements and environmental tests before their sale is permitted, to provide widespread education to users on proper application and safe use, to require certain specified standards of competence for those who wish to use pesticides, to control the quality, labelling and packaging of pesticides, to provide for the regular sampling and analysis of foodstuffs, to maintain a 'police force' of biologists and ecologists to detect misuse of pesticides and to keep a constant watch for adverse environmental effects and to enact regulations for the safe application of pesticides without danger to operators.

Governments should also sponsor and encourage investigations in Universities and Government-financed organisations into incidence and spread of pests and diseases, into the size of economic losses caused by pests and diseases in specific crops, and into the techniques of forecasting, as well as extensive basic research on the modes of action and fate of pesticides. Most Governments of developed countries do all or most of these things to a greater or lesser degree. The US Secretary of Agriculture has said that it is the policy of the Department of Agriculture to practise and encourage the use of those means of practicable, effective pest control which result in maximum protection against pests and the least potential hazard to man, his animals, wildlife and other components of the natural environment. Whether they should be financed from general taxation or partly directly from consumers by allowing food prices to rise is debatable, but the cost will certainly have to be borne by the public, as the price for receiving the benefits. Frisch says that, if a healthy environment is considered to be public property, it would be advisable to finance it by taxation in accordance with the rules of the financial sciences. Government must also ensure that manufacture of pesticides is a sufficiently rewarding investment to maintain the supply of crop protection agents which the nation needs. To this end, it may be desirable to lay down the methodology and protocol of minimum agreed and clearly specified toxicological and environmental tests which have to be carried out before sale of a pesticide is permitted and to arrange for any more extensive investigations which are deemed desirable to be undertaken by universities or Government-financed organisations. It may also be in the interests of the nation's agriculture to subsidise the users of pesticides, so as to make it possible for them to pay sufficient for solutions of minor crop problems or of minor localised pest or disease problems in major crops to make it economically attractive to manufacturers to seek and market appropriate products. Government should sponsor and encourage discovery and development of alternative methods of crop protection to use of pesticides, and of the manufacturing operations which are needed to make these available to growers on an adequate scale and at an acceptable price.

A very important area in which research and development is much needed is in formulation and application of pesticides. There are indications that changes in formulation may enable the same biological effects to be obtained with considerably reduced amounts of active ingredient. Methods of application of pesticides are crude compared with the precision with which drugs are administered and a large proportion of the pesticide never hits the target pest or disease but falls on the soil. Improvements in methods of application and in spraying equipment would make a very substantial contribution towards decreasing the total amount of pesticides used, optimising cost-benefit for the farmer, and reducing any risk there might be to wildlife or to the environment.

In total, Government should accept the concept of pest management in which use of pesticides is integrated with other methods of control of pests and diseases such as cultivational practices, development of resistant plant varieties, techniques of biological control, etc., in such a way as to optimise the cost-benefit picture for the whole community, and should sponsor whatever actions are needed to achieve this.

Although the social costs of pesticides are clearly very much less than those of the car and their benefits as great, or even greater, the public attitude to each of these products is markedly different. In their attitudes to pesticides, unlike their attitudes to the car, they are not aware of the extent of the benefits and therefore regard any risk, however small, as unjustified. This opinion is constantly fostered by the mass media. Government has a responsibility to promote education of the public to the facts of the case on a cost-benefit basis. It is clear that, if the public wants the benefits of pesticides with a minimum of risk, they will have to pay for them, which implies in effect, diversion of a significant proportion of the national product to this end, with concomitant sacrifice of an equivalent amount of consumer goods, and that to do this will have to be a conscious social decision. It is pertinent to remember that mankind has managed for millions of years without cars but could survive only a few weeks without food.

It should also be appreciated that production of food is an energy-intensive operation. Part of the energy comes directly from the sun through photosynthesis but the rest has to be supplied from exhaustible fossil fuel resources, for example, diesel fuel for tractors. Use of chemical herbicides as an alternative to traditional cultivating and weeding operations can often save significant amounts of such fuels. For example, it has been calculated that use of chemical instead of mechanical weeding in forestry can save about 13 gallons of diesel fuel per hectare per year. Another calculation has shown that direct drilling and minimum tillage techniques using the herbicide, paraquat, instead of conventional ploughing and harrowing can save about 7 gallons of diesel fuel per hectare.

The ways in which a Government should go about achieving the desirable control over use of crop protection and pest control chemicals is open to debate. In the USA, the tendency has been to rely on detailed and specific legislation, culminating in the US Federal Environmental Pesticide Control Act of 1972. In the UK reliance for safety has been placed entirely on voluntary schemes for co-operation between Government, growers, manufacturers and conservationists such as the Pesticides Safety Precaution Scheme and the Agricultural Chemicals Approval Scheme. The established common law of the UK restrains a farmer from allowing his crop protection operations to damage his neighbours' crops or livestock or to harm the public, and also forbids the sale of foodstuffs which can be proved unfit for human consumption because of the presence of contaminants. Apart from this, the only legislative control in the UK is the Agriculture (Poisonous Substances) Act which is designed to protect workers carrying out pesticide spraying operations. The voluntary nature of the UK controls over pesticides has led to co-operation and open communication between all interested parties and may well have resulted in more progress towards optimisation of cost-benefit than a system of legislative regulations. Principles for action which are agreed to be appropriate and proper by all who may be affected by them and which they all support are, in general, better from the community viewpoint than legislation which is imposed without full consultation and felt by some of the interested parties to be inappropriate and unjust. The Robens report in the UK on health and safety at work stresses the importance and value of consultations rather than regulations in ensuring safety.

The difficulties of carrying out a comprehensive cost-benefit study on a total community basis for any pesticide have already been stressed. The complexities are formidable and the amount of necessary data and the costs of obtaining it are very large. To complete a cost-benefit table as shown in Figure 2.1 (page 23) in its entirety may not be possible.

One of the great difficulties lies in the systems analysis aspect, that is, the interrelationships between various costs and benefits and the effects of change in one on all the others. Three examples may suffice. The withdrawal of DDT in the USA because of its alleged effects on the reproductive capacities of some predatory birds resulted in its widespread replacement in cotton-growing areas by methylparathion, as something had to be used to control boll-weevil if cotton production was to be maintained. Whereas DDT is safe to handle and virtually non-toxic to man, methylparathion is an acute poison and unfamiliarity with the necessary safety precautions has already led to many accidental deaths. The second example concerns introduction of the new high-yielding rice varieties into the Philippines where, as a result of fertiliser and pesticide use, greatly increased stocks of rice were accumulated which resulted in a vast increase in rat populations so that the extra rice in the event went to feed rodents rather than humans. The third example concerns the question of quantifying risk. If it is believed that toxicological and environmental studies carried out so far have reduced the risk from a particular pesticide to very small proportions, by how much would that risk be yet further reduced by carrying out, say, four times as much work? Would the increased benefit justify the costs?

Governments should not regard pest management in isolation but as an integral part of the system of food production for the nation. Apart from the specific actions suggested above, they should provide the finance, facilities and expertise in Government organisations for the overall cost-benefit and systems studies, to present informed and scientifically analysed data to the political decision takers. The problem for the electorate of any country of ensuring that the politicians to whom they have delegated their powers of decision act wisely in the interests of the nation as a whole on the basis of such data, is one for which the author does not consider himself qualified to suggest a solution.

9

Alternatives to Pesticides

If there is thought to be any risk that a particular pesticide might be harmful to the consumer, to the environment or to wildlife and non-target species, it is sensible to consider whether there are any effective and acceptable alternative methods of controlling the particular pest or disease. The word 'acceptable' is used because, when an alternative is contemplated, it is not sufficient to consider environmental effects in isolation. One must assess how the change to the alternative method will affect the whole cost-benefit situation. If the change introduces substantial new costs of various kinds, it must be decided whether these extra costs are sufficiently compensated for by a decrease in environmental risk. Possibly the alternative may be much more expensive than the original pesticide, in which case consideration has to be given to how these extra charges are to be met. Will they reduce the farmer's income, will they result in raised food prices, will they need to be financed by additional taxation? It will have to be asked what the consequences would be if the particular pest or disease was not controlled? Would this result in substantial losses of valuable foodstuffs? Apart from purely financial considerations the questions must also be asked whether the alternative will require more labour to apply, whether it will make more demands on the farmer's time, how it will affect the whole operation of the farm, etc.? Is the alternative one which the farmer can easily apply or does it involve his learning complicated new techniques or calling in specialists and experts to help him?

The alternative which will cause the least complications is to use another pesticide which has been proved to be free of residual or environmental effects or for which the risk of these is substantially less. In particular, it is desirable to replace all pesticides which are very persistent—that is, remain unchanged in the environment for long periods, with products which are only moderately persistent—that is, remain in sufficient quantities until they have done the job required of them and are then broken down to harmless substances by weathering or by attack from micro-organisms.

Apart from their use as insecticides, chemicals have long been used to repel insects rather than to kill them. Chemicals have also been used as attractants to lure insects into traps. Insects are believed to locate sources of food by following chemical stimuli—called phytomones—emitted from their food plants. Female insects often attract males of their species by emission of chemical stimuli called pheromones or 'sex attractants'. In some cases, it has been possible to isolate and identify these chemical stimuli and either manufacture them artificially or discover simpler and cheaper chemicals which produce the same effects. These can then be used to lure the insects into traps. Phytomones and pheromones are selective—that is, they affect only the target pest. Other novel chemical approaches are (1) to discover and manufacture chemicals which have the same effects as the natural hormones which control the processes of insect development, and to use these in the field to disrupt these processes in a pest population (hormones are, however, non-selective in their action), and (2) to discover chemicals which will inhibit the feeding response of the pest insects so that they effectively starve in the midst of plenty.

These novel chemicals are essentially 'second generation' pesticides. They could have advantages in that they would possibly be very active so that only small amounts would be required and that they would, except for hormones, be specific to the pest and be unlikely to harm non-target species or wildlife in general (although pheromones might attract closely-related species). However, few chemicals of these types have yet been discovered and the attempts to use those already discovered for insect control in the field have not been very successful. Triphenyltin compounds have had some success as Colorado beetle anti-feedants in Europe. Much more detailed knowledge of the habits and behaviour of target insects is needed if such chemicals are to be effectively utilised for commercial pest control in the future.

Novel chemical approaches to weed control which are being investigated include (1) use of substances which interrupt winter dormancy of weed seeds in the field so that they emerge at a time when climatic conditions are unfavourable for their survival, and (2) use of certain chemicals as seed-treatments to reduce the time between sowing and germination of crop seeds so that these emerge quickly and can compete more effectively with later-emerging weeds.

In some pest and disease situations there may be non-chemical alternatives to pesticides which are as effective but safer. Possibly in some situations such alternatives may also have additional advantages. The aim of crop protection and pest control research is to present the farmer with an extensive armoury for his continual battle against pests and diseases from which he can select the best weapons in each situation on a cost-benefit basis. To enable him to do this, it will be necessary to provide information on the various alternatives together with advice on how to use them and on their relative utilities and limitations. The responsibility for doing this will fall mainly on the advisory and extension services of Government agricultural departments assisted by other bodies such as farmers' unions, growers' associations and food processors.

The farmer may have, therefore, in any pest and disease situation, a range of options of which use of chemicals is merely one, albeit the most important and generally applicable one. These options need not, however, be mutually exclusive and the optimum choice from a cost-benefit viewpoint may be to use a number of them in combination or successively. In the remainder of this chapter very brief descriptions will be given of some non-chemical alternatives for the three main groups of pesticides—herbicides, fungicides and insecticides—and of some future possibilities. Those who wish for more detailed information should consult *Biology in Pest and Disease Control* (D. Price Jones and M. E. Solomon, Blackwell, Oxford, 1974).

9.1   Weed Control

Weeds have, in the past, been controlled by careful preparation of the seed-bed by ploughing and other cultivational practices, and by mechanical weeding and hand-hoeing. The latter operations are not possible in cereal crops for which, until the advent of herbicides, there were no effective methods for controlling weeds once they had emerged. Modern trends towards reductions in crop rotations, precision drilling, continuous cropping and high plant densities for many crops have made traditional methods of weed control difficult to apply, and there is scope for development of new techniques of cultivation.

Legislative control of the quality of crop seeds and regulations specifying maximum permitted contents of weed seeds in them have helped, in modern times, to reduce the spread of weeds in this way. Dissemination of air-borne seeds has been reduced by control of weeds in non-cropped areas, particularly by prevention of their seeding.

Biological control by introduction of insects which feed specifically on certain weeds has some applications. The most successful example is control of the prickly pear in Australia by the cochineal insect, *Dactylopius tomentosus*, and by the moth, *Cactoblastis cactorum*. This technique is applicable only to long-term control of a single dominant weed which is present over large areas of uncropped land. It is not suitable for rapid control of mixed weed infestations in particular crops or for the high level of control needed in arable crops. It is essential to be sure that the introduced insect will not attack related plants of economic importance nor produce any adverse effects on the natural ecological balance in the area. The potential environmental hazard of introducing such an agent might be much greater than that of any herbicide. (For a discussion of this complex subject the reader is referred to the article by Cussons in *Biology in Pest and Disease Control*.) A related technique which has had some success is the use of weed-eating fish, such as the grass carp, for control of aquatic weeds. This fish does not breed in the UK so could not cause ecological upset: it can be regarded as an environmentally safe persistent herbicide.

For the foreseeable future, weed control will depend mainly on herbicides, which play an essential part in modern intensive cultivation in developed countries. They are versatile, labour-saving, fuel-saving and convenient, and they give substantial economic returns for their cost. Their application by light tractors or from the air causes less disturbance to soil structure than compression by heavy tractors or continuous trampling by manual workers, and this can be an important factor in maintaining maximum fertility. In those very large areas of the world which are subject to wind or water erosion, use of herbicides to avoid ploughing and other massive disturbance of the soil can be of great value. For these reasons, herbicides comprise by far the largest proportion of all pesticides sold, and the demand for them is increasing rapidly. Fortunately, as has already been pointed out, they are generally not very poisonous and, as far as is known, they have no adverse effects on the environment and they do not induce resistance in the weed species. Possibly their overenthusiastic use might eliminate some potentially useful wild plants.

The trend for the future will be to regard weed control merely as a part of the total management of farm and regional agricultural systems. Development of this systems approach will involve balanced pro-grammes of rotation of crops, rotation of herbicides, use of combination and sequential treatments, and an integration of chemical control with cultural practices.

9.2   Disease Control

Until the introduction of fungicides and bactericides, mankind had been singularly unsuccessful in controlling plant diseases which were often the cause of total loss of crops and consequent famine. This crop loss still occurs in developing countries and, even as recently as 1970, 800 million bushels of corn were lost in the USA in an epidemic of Southern corn blight. This was an example of the damage which can be caused to susceptible varieties grown over large areas when weather conditions favour rapid spread of a virulent fungus. The total financial loss of about $2 thousand million was paid by the consumers of corn products since the price of corn rose from $1.28 per bushel to $1.60 per bushel as a result of the epidemic.

Build-up of populations of soil-borne diseases have, in the past, been prevented by systems of crop rotation, but there has been little direct protection against air-borne diseases. In modern times, there has been a greater appreciation of the importance of farm hygiene in limiting the spread of diseases. Removal of rubbish heaps, diseased tissue, dead leaves, fallen fruit, etc. which could be sources of infection is most desirable. It is interesting to note that, as long ago as 1873, Le Baron suggested that pigs should be allowed to run free in orchards to eat fallen fruit and thus prevent infection. Contaminated manure has often been a carrier of diseases. Legislative control over seed quality and restrictions on sale of infected seed have helped control infection, and plant quarantine laws have helped prevent spread of diseases from one country or state to another.

For many years it has been realised that external factors such as weather, soil condition, application of fertilisers, time of sowing, etc., could greatly affect the incidence of diseases. Thus, there is evidence that less damage is caused by yellow rust to wheat growing on heavy, deep, moist soils than to wheat growing on light, shallow, dry soils. An effect of soil aeration is suggested by the observation that red rot of sugar cane is common on impervious soils but rare on porous soils. Differences in soil drainage may account for variations in incidence of root rot in peas. It has been suggested that susceptibility to potato blight is increased by heavy nitrogenous manuring. Early-sown winter wheat is more prone to Helminthosporium disease than is a late-sown crop.

Studies of the life-cycles of fungi have revealed that some can live on alternative host-plants to the crops which they usually attack and that some need alternative hosts in certain stages of their development. Eradication of the wild host-plants may effect a significant degree of control. A well known example is wheat rust for which barberry is the alternate host. This connection was recognised long before the reason for it was understood, and a law was enacted in Massachusetts in 1760 for destruction of all barberry bushes.

There are very few examples—mostly in diseases of trees—of successful commercial control of fungal diseases by insects or other fungi. Apart from fungicides, the most important method for preventing loss of crops from diseases is by breeding resistant varieties of crop plants. This was first suggested in 1815 by Knight who advocated cultivation of fungi-resistant cereals. A practical difficulty is that it is not often that the variety which has the greatest resistance to disease turns out to give the greatest yields or best quality. Where this is the case, cost-benefit considerations may make it more advantageous to the farmer to grow the variety which gives the biggest yield and to protect it, if need be, by fungicides. It may take twelve years, or more, to develop a new variety for commercial use and a variety which has been bred during this time specifically for resistance to disease may have to compete with other new varieties bred during the same period for improved yield and quality.

In the past, resistance of a plant variety to a certain disease has generally depended on one gene. Sometimes this is sufficient to give a permanent solution to a particular disease problem. Resistance of potatoes to wart disease depends on a single genetic factor and the disease has been diminished in the UK by legal restrictions on cultivation of susceptible varieties. More usually, however, the disease develops new strains which break down the resistance of the plant, often quite quickly and suddenly. In most natural disease populations there are strains capable of overcoming inbred resistance in plants and it is, generally, only a matter of time before these strains become dominant. For instance, the yellow rust of wheat has continually developed new strains which have successively overcome the resistance of a range of new varieties. Resistance can also be broken down by external factors, e.g. if the plants are transferred to different soil or climatic conditions. This widespread failure of major gene resistance has led plant breeders to turn from using race-specific resistance ('vertical' resistance) to non-race-specific resistance ('horizontal' resistance). This approach coupled with developing knowledge of 'genetic engineering' offers more hope for the future, and plant breeding will certainly have an important future contribution to disease control. If this is to be achieved, it is necessary that as many varieties as possible of the major crops of the world and of their wild progenitors shall be collected and conserved in 'gene banks' to give the breeders sufficient basic material to work with.

The use of resistant plants is one of the most economical methods of disease control. Once a resistant variety which gives as good quantity and quality as susceptible varieties has been commercially established, there is no annual outgoing for the farmer for crop protection. However, resistance may break down after a time so a succession of resistant varieties may be needed.

However, as it is unlikely that a plant bred to be resistant to one particular disease will be immune to other pests and diseases, there are obvious limitations if the crop is likely to be attacked by a variety of these. In the process of natural selection varieties evolve which have a good average resistance, since the plant has to fight not only one but many enemies. It can happen that breeding of resistance to one particular disease increases susceptibility to other diseases. As with weeds, future management of diseases will depend on a judicious use of resistant varieties, cultural practices and fungicides in combination with each other and as part of a total system of farm management.

9.3   Insect Control

Traditionally, cultivational practices have been relied on for control of insects and nematodes, particularly the rotation of crops to prevent build-up of populations of soil pests, adjustments of sowing and harvest times, and removal of weeds, rubbish heaps, etc., which could be overwintering refuges. Thus, oats sown before the end of February generally escape damage by frit fly, whereas those sown in April are often severely attacked.

Breeding of varieties of crop plants resistant to or tolerant of certain pests has not been as well exploited or as successful as breeding for resistance to diseases, so there is considerable scope for future developments. It is believed that many plants, such as the yew and laurel, are naturally resistant to a variety of pests, because they contain natural repellant chemical substances. This offers the possibility of a new type of chemical approach to control.

Since it is mainly insecticides which have been suspected of producing undesirable environmental effects, it is in the field of insect control that most attention has recently been given to biological control, although this type of control was, in fact, used before insecticides became generally available.

One technique involves culture and release of millions of male insects which have been sterilised either by radiation or by treatment with suitable chemicals so that they compete with normal males for mates but fail to fertilise the females, so the pest population diminishes. World progress on application of this technique has been disappointing. One problem is the prodigious number of male insects required. It has been calculated that, if there is a natural population of one million virgin females in an area, then release of two million sterile males in each of four successive generations will theoretically reduce the population substantially to zero. This is obviously not a method which can be applied on a single farm, but must be applied to the total region in which the insect occurs, or to an area which can be geographically protected against an inflow from outside. Thus, it had a spectacular success in eradication of the screw-worm from the island of Curacao. The campaign against the Mexican screw-worm in the south-west USA necessitated release of 4 thousand million sterile males over an area of 85 500 square miles. It is a long term technique which cannot be used for rapid control of a sudden infestation. Also it is most effective at low pest population levels and cannot deal with major attacks. It is a technique which is really effective only with female insects which mate only once.

If this technique is to be used extensively in the future, economical methods of commercial mass-production of sterile insects will have to be developed. A deep understanding of the reproductive processes and habits of the insects which are to be controlled will also be needed. A major problem is that the insects to be sterilised usually have to be reared on the crop plant. It has been estimated that, in the Mediterranean area, rearing of sufficient olive fruit flies to provide effective control of the natural pest population would require a substantial proportion of all the olive trees in the area to be devoted to this purpose. It is therefore unlikely that the sterile male technique will be used on its own in the future. It is more likely that infestations will first be reduced to low levels by conventional insecticide treatments and then kept under control by sterile males.

A much more important method of biological control of insects depends on introducing a predator or parasite of the pest into the affected area. One of the earliest applications of this technique was the control in 1889 of the cottony cushion scale on citrus trees in California by the Australian ladybird beetle, *Vedalia cardinalis*. This method was highly successful. It has been suggested that birds might be used for biological control of insects, since they naturally devour large quantities of insects and insect larvae, but this idea has never been practically developed. An attempt was made in the USA to use bats to control codling moth but, by and large, control of insects has been by other insects. Most success has been with the use of imported predators or parasites to control imported pests, particularly in Australia, and difficulties are often encountered when attempts are made to control an indigenous pest.

The method has had some success in Southern Europe but none at all in Northern Europe. In the UK, the only examples of successful control have been in the highly controllable environment of the glasshouse. In the USA, up to 1956, 485 biological agents had been released against 77 species of pests. Of these agents, 95 became established and exerted some control over 22 of the pest species. Very large numbers of the predators or parasites are required. Thus, in Canada, between 1916 and 1956, one thousand million specimens of 220 species of parasites and predators were liberated against 68 pest species. Production of these numbers presents considerable difficulties, particularly if the parasite has to be reared on the host plants.

A related technique is the use of pathogens, that is, fungi, bacteria or viruses or their toxins which infect or poison the insect and cause its death. This is not an easy method to use since it is very susceptible to changes in external factors such as weather conditions. One of the most successfully used pathogens is *Bacillus thuringiensis*, which has proved effective against a number of species of caterpillar.

Control of insects by predators, parasites or pathogens can, if successful, be a cheap method of crop protection. Biological control agents also have the advantage of being highly specific in that they affect only the target pest and have no direct effects on non-target species. This specificity can, however, be a disadvantage when certain minor pests which are controlled incidentally by chemical pesticides used against major pests become major pests themselves when competition is removed by biological agents highly specific to the original major pests.

This is an example of Nature's tendency to fill any available ecological niche.

Use of biological control agents requires a great deal of care and extensive background studies. It is also a technique for use by specialist operators and not one which can easily be applied by the average farmer. It is applicable to large regions rather than individual farms and is a long-term measure which cannot be used to give rapid control of an unexpected infestation. It is not usually effective at low population densities of the pest, so it is normally necessary to maintain the pest population at a minimum level by artificial infestations to prevent the biological control agents from dying out. Expert judgement and constant observation are required to ensure that the right balance is maintained and this is very difficult to achieve in regions with unpredictable weather patterns.

Biological control is not without the possibility of environmental risk, since introduction of an alien life-form into an established environment is bound to have some effect on the equilibrium of species in that area. The possibility always exists that an introduced insect might become a pest of some economic crop. Thus, the larvae of blister beetles imported into the Philippines to control louse eggs gave adult insects which attacked lucerne. An introduced pathogen might change and become infectious to man or animals. It is therefore essential to undertake extensive studies before any new system of biological control is introduced, just as it is with a pesticide.

What then is the future for non-chemical methods of crop protection and pest control? Traditional methods based on cultivational practices, such as crop rotations and manipulation of sowing times, depths of planting, plant spacing, size of fields, sowing patterns, etc. will continue to be utilised and new methods such as direct drilling, minimal cultivation and modification of soil structure will be developed to meet the requirements of modern farming systems.

In temperate countries with rapidly changing weather patterns, such as the UK, pest and disease attacks on outdoor crops tend to be sporadic and rarely reach epidemic proportions. The instability of the environment tends to militate against use of sterile male and predator/parasite methods. The non-chemical method which has most future utility in these countries is development of resistant crop varieties. Plant breeders, aided by modern developments in genetic engineering will make an important contribution to future crop protection. In particular, the possibilities of plant breeding for resistance to pests, as distinct from diseases, have hardly begun to be exploited. However, this is a very long-term task.

In countries in which large areas of a single crop are grown, especially those in tropical or semi-tropical regions with regular weather patterns, sterile male, predator/parasite and pathogen methods may have utility provided that commercial mass production of the required organisms is feasible and that problems of registration requirements for use of predators, parasites and pathogens can be solved. It is implicit in such methods that they should be applied over whole regions, not individual farms, and so they will need to be applied by Government agricultural departments and financed on a national or regional basis from public money, as for public health or education. Use of such methods requires that pesticide usage over the region shall be fixed in advance and strict legislative control over use of pesticides by individual farmers in the region will have to be introduced. The modern farmer is working against a background of increasing land values and farm rents, high cost of marketing and fuel, shortage of labour and demands from the processed food industry for substantially undamaged crops. He will oppose any control measures which are labour intensive or which impose unacceptable constraints on his farming operations and he will wish to retain freedom to use the most economical methods to safeguard the quality of his crops. He considers that, if protection of the environment is a national interest, it should be paid for by the nation and that he should not be left, as at present, to foot the bill.

It seems clear that, within the constraints of the need for increased production of food, increasing costs of production and demand for high quality, the farmer will require crop protection and pest control methods which are cheap, highly effective, easily applied and quick acting and that chemical pesticides will be the mainstay for many years to come. However, all the non-chemical methods of crop protection and pest control will have their part to play, alongside pesticides, in the future. Pointless arguments over the respective merits or demerits of the various methods serve no purpose. What is needed is widespread adoption of the concept of pest and disease management by a carefully balanced use of all available methods in a way that is most effective against the pest or disease and least harmful to its natural enemies and to the environment. The decision on the precise way in which the various techniques should be investigated in each pest and disease situation obviously depends on a comprehensive cost-benefit study of that situation. The aim should be to keep pest populations continuously below a level at which unacceptable damage to the crop is produced—the so called 'economic threshold'. This level is not easy to define because there is a continuous relationship between pest numbers and crop yields which is influenced by environmental and climatic stress on both the pest and the plant. The fact that arable crops are short lived and nowadays often do not follow one another in set rotations may make a sustained policy of pest management difficult. A psychological barrier is the desire of the farmer to see an immediate and effective kill and a practical difficulty is the current demand of consumers and food processors for substantially complete freedom from blemish or damage.

It has already been pointed out that successful application of many non-chemical methods of crop protection and pest control demand a very detailed knowledge of the ecology, habits and life-cycles of the various pests and diseases. The amount of research work required is prodigious and it is pertinent to ask who will do it and how it is to be paid for? Although most large companies engaged on discovery and development of pesticides now devote considerable research resources to studying the problem of integration of chemical and non-chemical methods of crop protection, much of the basic knowledge required is more appropriate for academic than industrial research. The inescapable conclusion is that, if it is in the nation's interest to achieve efficient integration of chemical and non-chemical methods into total systems of pest and disease management, then the research work will have to paid for out of public money, that is, by the taxpayer and consumer.

A related problem for Government is that of disseminating to farmers all the detailed knowledge required for such systems of pest and disease management and of assisting them with their application, or of taking the responsibility for that application when this is needed over whole regions. Complex systems of pest and disease management would make far too great demands on the average farmer's time and technical knowledge. The main burden will fall on Government agricultural advisory and extension services, but it is likely that these will be augmented by emergence of a class of 'plant doctors' who will take full responsibility for the health of a farmer's crops. These may be independent consultants, comparable to veterinary practioners, or more likely, members of the staff of the larger pesticide manufacturers. It is logical that the pesticide industry should shift from being a purely manufacturing industry to being a service industry and that it should offer to its consumers a total pest and disease management service on a contract basis. From the grower's viewpoint, this would ensure safe operation and maximum financial return, from the food processors' viewpoint it would ensure maintenance of quality standards and defined crop protection programmes, from the consumer's point of view the danger of accidental misuse and introduction of toxic residues would be minimised and from the environmentalists' point of view all operations would be supervised by trained biologists and ecologists. For the manufacturer, it would ease the present economic difficulties of discovery and development of pesticides by giving him the added value of a service and this, in turn, would facilitate the introduction of a wider range of effective, environmentally safe pesticides for both major and minor crops by improving the financial expectations of such projects. It is possible that a comprehensive cost-benefit study of crop protection and pest control from the point of view of the community as a whole may point to a move in this direction as an approach to optimisation. Many of the larger pesticide manufacturers have moved a long way in this direction by providing extensive expert technical service to farmers, although their representatives may need to acquire considerably more biological expertise in order to cope with future pest management programmes.

Public and Animal Health and Related Topics

The pesticides aspects of public and animal health comprise:

1. Control of biting, irritant, noxious, annoying or contaminating insects, mites and vermin in homes, factories and public buildings, especially those in which food is stored, handled or prepared, in animal and livestock houses, on and around refuse tips, agricultural waste heaps, etc., in and around drains, sewers, etc., and in recreational areas.
2. Control of insects and mites which infest humans and animals as parasites.
3. Control of blood-sucking and biting insects and mites which carry the germs of disease which infect humans and animals.

Insects, mites and vermin of type (1) are pests in all countries of the World, though to a greater degree in tropical areas. Control of flies, cockroaches, rodents, etc., in places where food is stored, handled or prepared is essential to prevent bacterial contamination of foodstuffs.

World losses of food crops during transport and storage after harvest are prodigious. Part of these losses is caused mechanically by careless packaging or handling but a considerable proportion is caused by fungi, bacteria, insects, mites or rodents which make whole cargoes or batches of foodstuffs unfit for human consumption. Table 10.1 shows an estimate by the Food and Agricultural Organisation of the United Nations of the total percentage post-harvest losses of a number of crops under tropical conditions. In all a highly conservative estimate is that around 30 million tonnes of food produced in tropical areas are lost each year by post-harvest wastage. Even in the temperate developed countries with all their modern aids post-harvest spoilage can be very substantial, for example, it is estimated that about 25% of the total citrus crop in the USA is lost in this way.

On an investment appraisal basis the cost of chemical agents to control the various pests and diseases which attack crops during transport and storage can be set against the value of the foodstuffs saved. However, as most consignments of foodstuffs are insured, there is little inclination at the moment for shippers or wholesalers to take an interest in the problem, although insurance companies are beginning to show some concern. From a cost-benefit viewpoint it is obvious that pesticides which are used to treat actual foodstuffs prior to consumption must be chosen very carefully to have very low toxicities and not to leave significant amounts of residues.

When pesticides are used in any of the other situations in (1) the investment appraisal approach does not apply, since, although there is a cash input in terms of cost of the treatment, there is no direct cash gain and any benefits are social benefits or benefits to health.

In the case of recreational areas such as beaches, parks or picnic areas, the public does not want to be pestered by flies, wasps, gnats or similar insects. As tourism is part of the economic well-being of many regions there is a direct financial incentive to institute programmes of pest control in order to make the amenities more attractive to visitors. Furthermore, such programmes may include the management of recreational areas by, for example, removing plants which are unsightly or which can cause injury, such as brambles, poison ivy, etc. The untamed jungle is not a very pleasant place for a picnic and the public by and large would not welcome nature in the raw, but demand the aesthetic pleasures of the countryside without any of its natural inconveniences. The same considerations apply to the management of recreational water, for example, to prevent rampant growth of aquatic weeds. A cost-benefit study would have to balance the costs of chemical control agents against the gain to the regional economy from tourists, and particular thought would need to be given to possible unwanted environmental and ecological effects of management programmes covering large areas of land or water.

TABLE 10.1
Post-harvest crop losses under tropical conditions. (From D. G. Coursey, *Biodeterioration of Materials*, Vol. 2, Applied Science Publishers Limited, 1972.)

Commodity	% Loss
Avocados	43
Aubergines	27
Bananas	33
Cabbages	37
Carrots	44
Cauliflowers	49
Grain	25
Lettuces	62
Mangoes	30
Onions	16
Oranges	26
Pineapples	70
Potatoes	8
Sweet potatoes	95
Tomatoes	30
Yams	15

The insects and mites of type (2) have much more serious effects on human and animal health as distinct from human enjoyment. Bedbugs, lice, fleas and other pests have debilitating effects on people in temperate zones, and, in tropical areas this list is augmented by a variety of parasites which can cause severe sickness and death and which bring suffering and misery to millions of people.

Parasites of animals are often hideous in their effects and, besides causing suffering to the infected beasts, have an adverse effect on the economics of livestock production and reduce world supplies of much-needed animal protein (see Plates 5 and 6). Amongst major pests of animals mention can be made of ticks on cattle, warble fly and blowfly on sheep, lice on animals of all kinds and mites on poultry. Use of veterinary pesticides can be assessed both from an investment appraisal viewpoint, in terms of greater numbers of animals surviving, bigger weight gains and more efficient conversion of expensive foodstuffs, or from a cost-benefit viewpoint which will take into account the well-being of the animals and, as with pesticides applied directly to foodstuffs, will consider the risks of residues in meat and dairy products.

Insects and mites of type (3) which feed on the blood of humans and animals are, on a total world basis, the most severe public and animal health problem. Although in most developed countries the danger has been kept largely under control by public sanitation and hygiene measures there are within them semi-tropical areas, as, for example, Florida in the USA, where disease-bearing insects are indigenous and have had to be controlled by chemical means. It is in the tropical areas of the developing countries that the magnitude of the problem reaches staggering proportions in terms of human suffering and death. Malaria, yellow fever, dengue, encephalitis, filiariasis, trypanosomiasis, onchocerciasis, leishmaniasis comprise but a few of the wide range of destructive diseases which are carried by insects to man and his animals under these conditions (see Plates 1, 2, 3 and 4). Assessment of control measures is entirely a matter of cost-benefit studies because there is no direct financial gain. The problem of putting a value on human life and health is difficult and is not made easier by widely diverging opinions on this subject and by the varying attitudes of the public in developed countries to the populations of the developing countries. The chronic debilitating effects of endemic diseases make it impossible for such countries to raise their economic statures.

One example must suffice because of the constraints of space in this short book—namely, that of malaria which is notorious for the extraordinary amount of mortality and morbidity it causes. About one thousand million people in the world are estimated to be at risk from this disease which is transmitted by mosquitoes. The only effective and economical method which has been discovered to bring the disease under control is by widespread application of insecticides. The possible alternative of improvement and sanitisation of the local environment is prohibitively expensive for most developing countries where ill-health and poverty form a vicious circle.

During the past twenty years a global eradication campaign has been carried out under the auspices of the World Health Organisation in 124 countries with a total population of 1724 million. In 19 of these, malaria has been completely controlled and in the rest it has been substantially reduced as shown by representative figures in Table 10.2.

TABLE 10.2

Effects of WHO malaria eradication programme. (From *The Place of DDT in Operations against Malaria*, World Health Organisation, Geneva, 1971.)

Country	Year	Number of malaria cases
Mauritius	1948	46 395
	1969	17
Cuba	1962	3 519
	1969	3
Dominica	1950	1 825
	1969	Nil
Dominican Republic	1950	17 310
	1968	21
Grenada	1951	3 223
	1969	Nil
Jamaica	1954	4 417
	1969	Nil
Trinidad	1950	5 098
	1969	5
Venezuela	1943	817 115
	1958	800
India	1935	100 000 000
	1969	286 962
Sri Lanka	1946	2 800 000
	1961	110
Bulgaria	1946	144 631
	1969	10
Italy	1945	411 602
	1968	37
Romania	1948	338 198
	1969	4
Spain	1950	19 644
	1969	28
Taiwan	1945	1 000 000
	1969	9
Turkey	1950	1 118 969
	1969	2 173
Yugoslavia	1937	169 545
	1969	15

Although, as has already been observed, it is difficult to assess the social benefits from such reductions in ill-health, there are, nevertheless, some measurable economic effects which result from increased production of the available labour force, reduction in the high costs of medical treatment, and increases in land values in areas where only a subsistence agriculture could previously be carried out. It has been estimated that continuance of malaria in Syria at the 1950 level would have completely frustrated the entire development effort of the last decade in that country by its effect on labour supply through loss of working time, reduced efficiency and higher death rates.

The insecticide on which the malaria eradication programme has depended, and still depends, is DDT. At the time of maximum use, over 60 000 tonnes per year of DDT were utilised, mostly in circumstances involving human contact. The interior walls of millions of houses and huts were sprayed with it and it was even occasionally added to drinking water and sprayed on edible plants and animals. In all this prolonged experience there have been no reported instances of adverse effects in any human being and there is a substantial body of evidence to suggest that DDT is one of the least toxic and safest of all pesticides to man and animals. However, DDT is an extremely persistent chemical which is not rapidly broken down under natural conditions, in fact, its usefulness in the malaria eradication programme depends on this property. It is believed that the presence of DDT in the environment during the past thirty years has caused reduction in fishery products from lakes, streams and off-shore areas and has interfered with the reproductive capacity of certain species of predatory and fish-eating birds, although there is no firm evidence that this is so. Because of this, use of DDT has been severely curtailed in the USA and other countries mainly as a result of extensive political pressures. These pressures are now being directed towards the objective of a global ban on manufacture and use of DDT imposed by the political and economic power of the developed countries. There is sympathy for this action amongst some politicians in some developing countries who, unmindful of the different circumstances and conditions of their countries, react on the basis that it is wrong for coloured people to be exposed to a risk which white people will not accept.

What would be the effect if DDT was withdrawn from the WHO anti-malarial programme? In Sri Lanka, where this happened, the number of malaria cases rose from 110 in 1961 to 2·5 million in 1968–9. Malaria cannot be conquered: it can only be kept at bay. The only available insecticides which might conceivably be effective replacements for DDT are much more expensive, the cheapest being three times the price. Unfortunately, these possible replacement materials are much more poisonous to man than DDT and have to be applied much more frequently, so that extensive personnel, organisational and logistic changes would have to be made in the anti-malarial campaign, such as direct medical supervision of spraying, use of protective clothing, strengthening of supervision, intense and frequent training, etc., which would greatly increase the cost of that programme. Are the affluent nations prepared to meet this extra cost, which would amount to about $300 million if the cheapest replacement for DDT was used?

Biological control by larvivorous fish, microsporidia, fungi and bacteria have been investigated, but can be used only in special circumstances and are a practical possibility only for the very distant future. Such methods may not be entirely without risk for man, because some of the organisms secrete toxic metabolites into water and also because of the potential dangers of pathogenicity if new micro-organisms are introduced into an environment. Very detailed safety studies would therefore have to be made before they could be used. Genetic engineering as described by Davidson is also a future possibility.

Since DDT is used in the malaria eradication programme almost exclusively indoors as a wall paint, there is little risk of serious contamination of the external environment. On a cost-benefit basis the case for its continued use would seem to be overwhelming. The developed countries should not only discourage restrictions on use of DDT for malarial control but should ensure that adequate supplies are available as cheaply as possible to the developing countries. Resolutions to this effect were passed at the 22nd session of the WHO Regional Committee for South-East Asia in 1964 and by the 23rd World Health Assembly in 1970. The WHO has stressed that withdrawal of DDT would be a major tragedy in the chapter of human health and would condemn vast populations to the frightening ravages of endemic and epidemic malaria.

It is desirable that outdoor use of DDT, especially near watercourses, rivers or lakes should be discontinued except where there is no alternative. In fact, it is not now used by the WHO for outdoor control of any insect-borne tropical diseases except to a limited extent against trypanosomiasis in Africa. Effective larvicides and imagocides are available for control of the culicine mosquitoes which carry filariasis and encephalitis and the aedine mosquitoes which carry yellow fever and dengue. Methoxychlor is now used against the blackfly vectors (i.e. carriers) of onchoceraciasis and lindane, malathion, and carbaryl against the rat flea which carries plague and the body louse which carries typhus.

The above discussion of the role of pesticides in human and animal disease control in developing tropical countries leads naturally on to the question of the total pesticide needs of these countries. This is a subject which deserves a book to itself and cannot be adequately dealt with within the space restrictions of this book, which has concentrated almost entirely on crop protection and pest control in the developed countries. Nevertheless, the author considers that it must be mentioned, even if only superficially. The difference in circumstances between the peoples of the developed countries and those of the developing countries is staggering, and the cost-benefit analyses of pesticide use are therefore quite different in the two environments. The affluent well-fed inhabitants of the developed countries can afford the cost of discovery, development and use of crop protection and pest control chemical or non-chemical methods which reduce the risk of toxic residues or environmental hazards to a vanishingly small level. They can also afford to set tolerance limits for residues or other regulatory restrictions on certain pesticides even if this decreases the yield of some crops, because adequate alternative types of foodstuffs are available to them. For vast populations in developing countries living in poverty and often dependent entirely on one staple crop the situation is quite different, and it may be expedient for them to accept a higher degree of risk of toxic residues or damage to wildlife because of the much greater risk of starvation. This point has been made repeatedly by Dr J. M. Barnes, the Director of the Toxicology Research Unit of the UK Medical Research Council. Such populations can afford only the cheapest pesticides and these include some which have come under criticism in developed countries with regard to residues and environmental hazards. The tendency of bodies such as the Food and Agricultural Organisation to lay down residue tolerances for all pesticides for acceptance on a world wide basis may have disastrous implications if applied to the developing countries, since the tolerances recommended are invariably those which are acceptable to the developed countries should not only discourage restrictions on use of DDT for appropriate for the developing countries does not imply that those populations should be exposed to real danger because, as has been pointed out in Chapter 6, the tolerances specified by regulating authorities in developed countries have extremely large built-in safety margins. The decision is related to the cost-benefit situation in any particular area especially with regard to the amount of risk which is justified by the benefits. It is important to realise that crops such as cocoa, coffee and sugar are the main basis of the economies of many of the developing countries and the only means by which they can acquire foreign exchange for industrial improvement.

Another problem which becomes acute in tropical areas is biodeterioration of materials. Although the destruction or spoilage of timber, textile materials, and other fibre products, paint work, plastics, rubber, antiques and objects of art are substantial in temperate countries and cause very great economic losses, it is under hot, humid conditions that destruction and spoilage are exacerbated and have profound effects on the economic viability of communities. The losses of crops after harvest have already been referred to earlier in this chapter.

However, it is food production which is by far the most pressing problem in developing countries in the face of rapidly increasing populations. By the year 2000, 80% of the world's population will live in developing areas. It is, therefore, as an aspect of the total problem of food production in these countries that use of pesticides must be primarily considered. Along with pesticides, these countries need improved seed, adequate supplies of fertiliser, better cultivation equipment and extended irrigation schemes. It has been estimated that by 1980 world population will have risen to 43% above the 1963 level and that to increase food production by this amount will require capital investment of $12 thousand million of which 8% will be for pesticides, 1% for seed and 91% fertilisers. Banking institutions or Governments will have to extend credit for $2·9 thousand million to farmers to buy these materials. To provide the estimated need of about 600 000 tonnes of pesticides will require 187 new manufacturing facilities, 332 formulating plants and 1·1 million square metres storage space around the world. The relationships between crop yields and pesticide usage over an eleven-year period are shown in Figure 10.1.

It is questionable whether the right course is to introduce highly sophisticated chemical manufacture and chemical products, which are designed for the capital-intensive, labour-saving economies of developed countries into subsistence agriculture which has no capital but plentiful labour. The developments which have taken hundreds of years to mature in the developed countries cannot be introduced in totality into the developing countries overnight. Possibly the correct approach is by way of the so called 'intermediate technology' in which

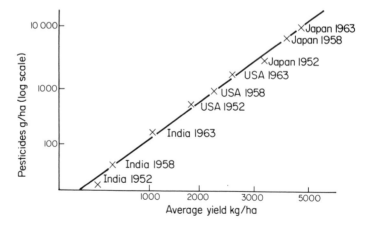

FIGURE 10.1   Relationship between crop yields and pesticide usage over an eleven-year period. (By courtesy of W. B. Ennis, United States Department of Agriculture, Beltsville, Maryland.)

improved hand implements are preferred to tractors, simple irrigation schemes to massive dams and use of indigenous local materials is preferred for crop protection and pest control. With regard to the latter point, it is interesting to note that it has recently been discovered that effective insect repellants and insecticides can be isolated from the Neem tree and from the Chinaberry, both of which grow in profusion in India and South America. Nevertheless, for the immediate future, the developing countries will need large quantities of cheap pesticides to be manufactured in the developed countries and supplied to them extremely cheaply, and these will include pesticides which are environmentally suspect in those countries. Will the developed countries be willing to meet the tremendous costs of producing these pesticides, together with fertilisers, seeds, equipment and irrigation schemes? The alternative may be, as C. P. Snow has said, to watch on our television sets the spectacle of millions of people in less privileged countries starving to death before our eyes.

The author dares to hope that one day the nations of the world will unite to solve the food problems of the world, that Governments, universities, manufacturers, growers, environmentalists and conservationists will then work together to prevent irreparable harm to the species *Homo sapiens*, that crop protection and pest control chemicals will then take their rightful and respected place as one of the means to this end and that optimisation of cost-benefit of pesticides can then be approached on a global basis. A necessary prerequisite is that the public at large in all countries should understand and assess these costs and benefits in a rational manner free from emotion or hysteria, and it is as a small contribution to this aim that this book has been written.

Time is not on our side. As a result of the drought in 1972 North American grain stocks are exhausted. The whole world now depends on each year's harvest for next year's food. Yet population growth, particularly in developing countries, continues at an alarming rate. Starvation may become, for the developed countries, a present reality rather than a distant phenomenon.

Recommended Further Reading

(The author gratefully acknowledges all these books as source material for this publication.)

The Economy of Cities, J. Jacobs, Cape, London, 1970

Pests of Field Crops, F. G. W. and M. G. Jones, Arnold, London, 1974

Losses in Agriculture, US Dept of Agriculture, Agricultural Research Service, Agr. Handbook No. 291, 1965

Untaken Harvest, G. Ordish, Constable, London, 1952

Mankind and Civilisation at Another Crossroad, N. Borlung, Food and Agriculture Organisation, Rome, 1971

Chemicals for Pest Control, G. S. Hartley and T. F. West, Pergamon Press, Oxford, 1969

Scientific Principles of Plant Protection, H. Martin, Arnold, London, 1928 and subsequent editions

Restricting the use of Phenoxy Herbicides, US Dept. Agriculture, Agricultural Research Service, Economic Report No. 194, Washington, 1970

Pesticide Industry Profile Study, National Agricultural Chemicals Association, Washington, 1971

Research on Pesticides. C. O. Chichester, Academic Press, New York, 1965

Scientific Aspects of Pest Control, Natl. Acad. Sci. Natl. Res. Council Publ. No. 1402, Washington, 1966

Chemical Fallout: Current Research on Pesticides, N. W. Millar and G. C. Berg, C. C. Thomas, Springfield, 1969

Pesticides and Pollution, K. Mellanby, Collins, London, 1967

Silent Spring, R. Carson, Houghton-Mifflin, Boston, 1962

Agriculture and the Quality of our Environment, N. C. Brady, Am. Assoc. Adv. Sci. Publ. No. 85, Washington, 1967

Pesticides in the Environment and their Effect on Wildlife, N. W. Moore, Blackwell, Oxford, 1966

Ecology and the Industrial Society, R. W. Edwards and J. M. Lambert, Wiley, New York, 1965

Report of the Secretary's Commission on Pesticides and their Relationship to Environment Health, US Dept. Health, Education and Welfare, Washington, 1969

Homo Sapiens—the Species the Conservationists Forgot, D. C. Hessayon, Chemistry and Industry, London, 1972, pp 407–11

Pesticides: A Code of Conduct, British Agrochemicals Association, London, 1968

Food and Health: The Safe Use of Pesticides, J. M. Barnes, British Food Journal, Uplands Press, Croydon, May 1967

Third Report of Research Committee on Toxic Chemicals, Agricultural Research Council, London, 1970

Review of Organochlorine Pesticides used in Great Britain, HMSO, London, 1969

Toxicants Occurring Naturally in Foods, Natl. Acad. Sci. Publ. No. 1354, Washington, 1966

The Place of DDT in Operations against Malaria, World Health Organisation Record No. 190, Geneva, 1971

The Pesticide Problem: An Economic Approach to Public Policy, J. C. Headley and J. N. Lewis, Resources for the Future, Washington, 1967

Technological Economics of Crop Protection and Pest Control, Society of Chemical Industry Monograph No. 36, London, 1970

Social and Economic Values in the Assessment of Crop Protection and Pest Control Methods, Society of Chemical Industry Monograph, London, 1975

Index